SUISUN MARSH HISTORY

SAN FRANCISCO AND SUISUN

(Approximate Scale: 1" to 12 miles)

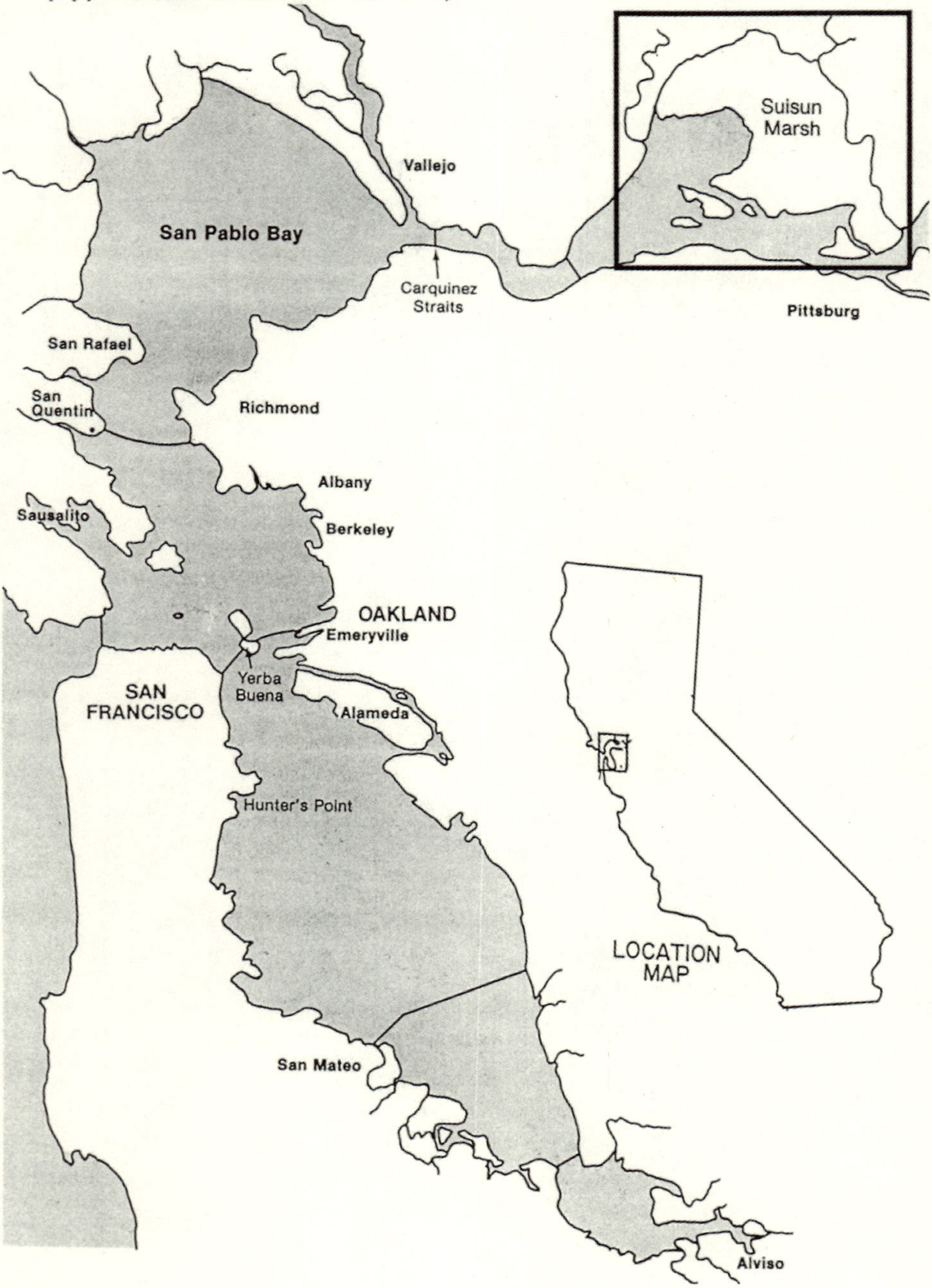

Suisun Marsh History

Hunting and Saving a Wetland

By

Anthony Arnold

Illustrations by Rich Radigonda and Boris Ilyin

Address orders or inquiries to

Anthony Arnold
P.O. Box 1252
Novato, CA 94948.

ISBN 1-880710-04-8

Published by
Monterey Pacific Publishing Co.
P. O. Box 829
Marina, CA 93933

This book is dedicated to the memory of George Stanleigh Arnold (1881-1942), and to all those others, living and dead, who have contributed to the preservation of the Suisun Marsh as wildlife habitat.

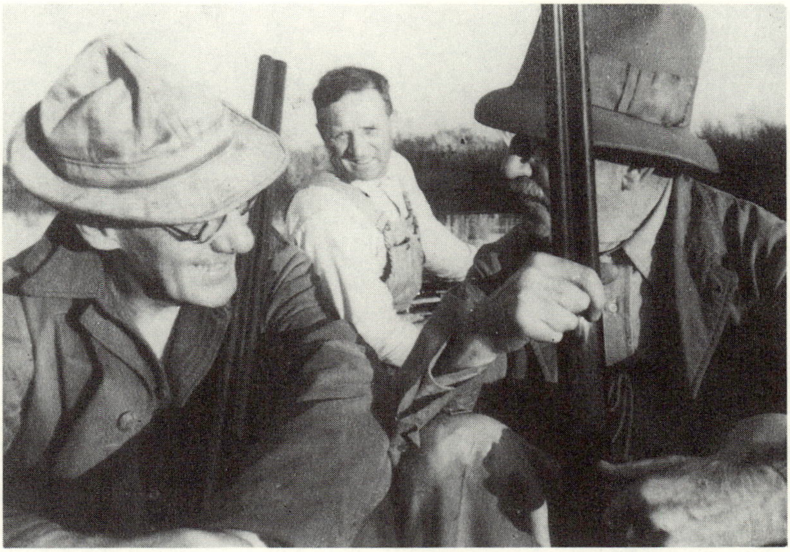

Photo by Peter Arnold

Big Stan, Gus, and Mr. Hopps (an Arnold Club Regular) ca: 1940

CONTENTS

November, 1940

Since before dawn the boy has been alone in the calm rain. His boots are leaky hand-me-downs that earlier, when he waded to the blind in the dark, let icy marsh water into one leg. Most of the water has now leaked out again, but his hat and outsized jacket are sponging up moisture from the top down. Beside him, the unpedigreed, single-shot, 20-gauge hammer gun is depositing its beaded drops on his hand and thence up his sleeve. In the boy's whole body, only the one dry foot is comfortable and warm.

The boy is too short for the blind. Perched as high as he can sit, he can see only the grey sky above and the concealing pickleweed that brushes his face. Without standing, he can only hear, not see, the thorough, unhurried rain spickling the pond. Nothing is flying.

Then a flicker of movement overhead, and dropping straight down on him with set wings is a duck. It seems much too small, impossibly high, but the boy cannot help himself. He flounders to get the gun stock past the folds of his jacket, cocks the hammer, and fires blindly, sure he will miss as the gun barrel shuts off his view of the bird. But as the gun lowers he watches in disbelief while the duck, a hen sprig with wings now tucked into her body, hurtles down to splash him as she lands, dead, not three feet away.

He will never forget that morning on the Suisun Marsh, that moment which, more than fifty years later, is the cornerstone of this book.

Acknowledgments

Never has one owed so much to so many. The following is an incomplete listing of some major contributors:

First the institutions and their people: Solano County Archives (Jerry Bowen, Bert Hughes, Judy Nielsen, and everyone else there); Solano County Recorder's Office; California Academy of Sciences; California Waterfowl Association (Becky Easter and others); San Francisco Maritime Museum; the UC Berkeley's Bancroft and other Libraries; Department of Fish and Game (Chuck Munroe, Dennis Becker, and many others); Suisun Resource Conservation District (Toni Dynan, Lee Lehman, Greg Martinelli, Steve Chappell, et al.); Western Railway Museum (Bart Nadeau); Vacaville Museum (Delia Irving and all the rest); Library of Congress (Dave Kelly); plus all the local public libraries and their staffs, from Fairfield to San Francisco to Marin.

As for individuals, well, we can start with Peter Avenali, Jim Bancroft, Colleen and Kenny Border, Dan Chapin, Kris Conti, Bill Coon, Ed Cutter, Jean Davisson, Stan Dickover, Dick Dinkelspiel, Mel Frohrib, Bill and Joan Frost, the late Ed Gillette, Dick Griffith, Bernice Huber, Les Jacobson, Frank Johnson, Jack Keeler, Don Kibby, Jim Koeppen, Bill Langbehn, Ray Lewis, Ike Livermore, Clyde Low, Martha McGettigan, Walt Powell, Rich Radigonda, Roy Saucerman, L. R. Skinner, the late Allan Steinau, Rick and Diane Tesene, and Ted Westphal. There were many, many more. Apologies to those I have left out.

And finally to the two best brothers, Stan and Pete Arnold, and the one best wife a man ever had. The first two inspired this book, and the last, Ruth, not only did much of the research and editing but endured an ever-more grouchy spouse in the process.

None of these helpers gets me off the hook for any mistakes. My apologies especially to Clyde Low, whose objections I was able only partially to sustain when dealing briefly with the Suisun branch of the Patwin Native Americans.

PREFACE

The Teal Duck Club lies on the western side of the Suisun Marsh, just across the Southern Pacific tracks from where I shot my first duck in 1940. Its fireplace, in front of which I took my initial notes for this book, has two sets of andirons: one made from old railroad rails that must have been cut by a torch, fired to white heat, and bent through 90 degrees to get the proper shape; and the other from upper-class brass, crafted originally for the carriage trade but now tarnished and sooted beyond recognition. They are equally effective in supporting the snapping logs that warm our living room, a job they share with the heat-warped, middle-class grate that lies between them, all three cooperating with classless ease.

To me, the andirons are like the inner core of the Suisun duck-hunting community, a mixture of people spanning a vast range of incomes and from all walks of life, mostly men who have little in common except an intense love of the marsh and of hunting. Like the unpolished old brass andirons, the richest are reduced by their stinks, tattered clothes, and general decrepitude to the level of the poorest. (Never mind how they leave home— "One short hour at the marsh suffices to transform the dapper hunting swell into a dirty old hippopotamus," that from *The Breeder and Sportsman* of 12 December 1885.) Their hunting peers have always judged them neither by outward appearances nor by how much money they have but by a whole range of other values they bring to the marsh.

Beyond the inner core there are others for whom the marsh and the hunting are only occasional amusements, to be forgotten between seasons and even between weekends—casual visitors, as a bird watcher might call them. They do other and, to them, sometimes more important things, from watching football games to fishing to skiing to playing golf. Some of them are excellent shots, but that has little to do with it. The proficient are not always the dedicated, nor the dedicated proficient.

Some might liken the inner core to a cult, but that does not do us justice. Cults tend to be tightly disciplined, conformist groups, feeding on their need for mutual support. By contrast duck hunters are independent individualists, but they are no less devoutly committed, in their own peculiar way, than regular church-goers.

My father—Big Stan to his offspring—provided a good illustration of this devotion in the 1930s. Our minister, John Leffler, had approached him to teach Sunday school and he had agreed. Not only that, he pledged to see to it that his four sons attended. When this news broke, it caused instant consternation. In those days, you could only shoot on Wednesdays and Sundays, and school already eliminated Wednesdays for us boys. But our father stood firm until our brother Pop had an inspiration:"But Big Stan, if *we* can't hunt, neither can *you*." Silence. Then, after dinner, an overheard muffled telephone conversation: "Uh, John, I suddenly realize I'll be tied up on Sundays until aft.. until mid-January."

Of necessity this book just skims the surface. There are about 153 clubs on the Suisun Marsh, many of which have histories that could take full book-length coverage. (My own research has resulted in a data base of more than a thousand entries for the early hunters and clubs, there is lots more waiting in my file folders, and yet more remains unexploited in various archives. If anyone wants help in fleshing out his own club's history, please let me know.) Just the factors that have saved the marsh from "development," ranging from blind luck to agricultural economics to dogged political battles, could fill shelves. To report on all this in detail in one book would produce something too big in size and heavy in content for anyone to read. Instead, I have tried to focus first on the old history, before any of us was alive, and then to boil down the essence of living memory, with emphasis on personal stories and observations. If I have concentrated on the northwest corner of the marsh, and specifically on the

Teal and Cordelia clubs, it is because this is the area where it all began, and in the early days the Teal and Cordelia caught a lot more journalistic attention than any of the others.

Appendixes A and B have little to do with history but are intended to give a nostalgia trip to retired marsh rats and an explanation to strangers of our infatuation with this unique wetland.

The purpose of the book, aside from my own selfish indulgence and hopeless addiction to research, is the hope that it might contribute, however marginally, to preservation of the marsh and its hunting tradition.

Finally, I ask that anyone with more or better information than I have provided share it with me. If nothing else, I want to leave behind a better set of archives.

Tony Arnold
Novato, 1996

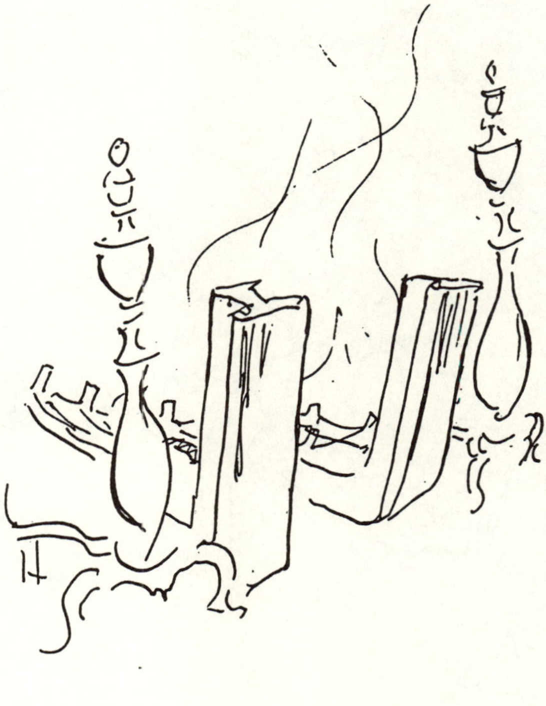

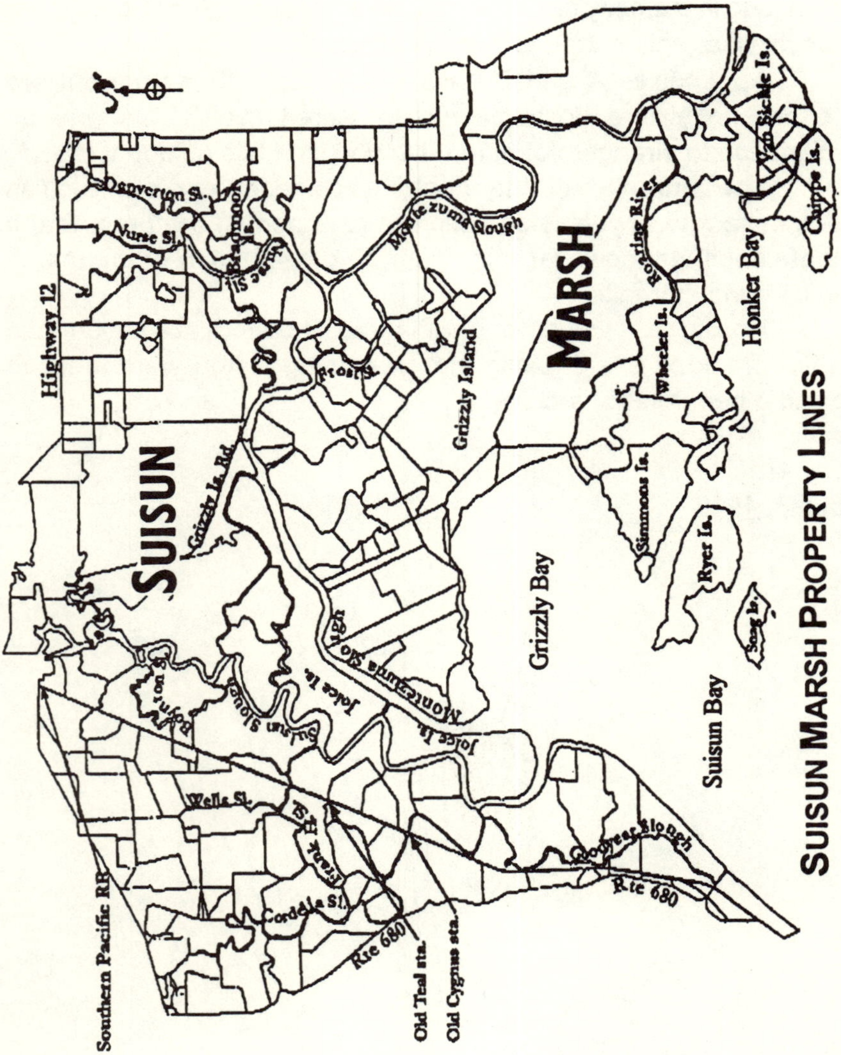

SUISUN MARSH PROPERTY LINES

CHAPTER 1

CREATION AND EARLY SURVIVAL

Geologic History

In the sweep of geologic history, the Suisun Marsh is scarcely more than a blink. A mere 10,000 years ago, toward the end of the last ice age, there was no marsh, but only the lower reaches of a great river, still 300 feet above tideline. The seashore was off beyond the Farallon Islands, more than 30 miles west of the Golden Gate.

If you take the whole span of the earth's history—some 4-5 billion years—as the height of one of the 746-foot Golden Gate Bridge towers, then those 10,000 years are less than the thickness of the pages you have turned so far in this book. Yet that tick of the geologic clock has seen profound changes in the Coast Range gap through which the water flowed—and still flows—to the ocean.

The Bay Area's soft hills and two main peaks, Mt. Tamalpais and Mt. Diablo, are largely unchanged, but in place of today's tidal flats, bared and flooded every twelve hours, there were steep canyons. These carried the last feeder streams to the river that had already linked the two major central-valley systems, the south-flowing Sacramento and the north-flowing San Joaquin. From the Klamath Mountains along the Oregon border to the Tehachapis in the south, the two rivers flowed toward each other to meet briefly before turning west to the sea. It was not a world-class system—not a Nile or

Amazon or Volga or Mississippi—but its 150-mile east-west by 500-mile north-south basin could have accommodated England and Wales with room to spare.

From the standpoint of the rivers' influence on the land, the main difference between yesterday and today lies in the speed of their flow and their ability to carry sediment. Only after the ice caps melted enough to raise the world's sea levels by 300 feet and drown the canyon that is today San Francisco Bay did the currents slow enough to begin releasing their silt. These deposits slowly built up almost to sea level, spreading sideways up the tributaries in the breach the river forced through the hills in its fast-moving youth. Back upstream, beyond where the San Joaquin and Sacramento join, the rivers began to meander aimlessly, building the delta as the Indians and first white men knew it. As fresh water from the mountains met the salt water from the sea in an ever-changing mixture, a uniquely rich habitat developed on top of the deepening layers of sediment and peat. Again, it was a geologically swift action; recent corings have shown that some of the deep marshlands are only 2,000-3,000 years old.

Before diking and draining began in the 19th century, the delta spread across more than 750,000 acres. Now it is virtually all enclosed and managed, and most of its eastern wetlands have long since disappeared under the plow or bulldozer. In the west, however, lies the Suisun Marsh, which also lies behind six- to eight-foot levees but is maintained mostly as wildlife habitat. Originally only a tenth of the whole delta, today it is the largest contiguous estuarine marsh in the lower 48 states, supporting 200 species of birds, 45 kinds of mammals, 36 different reptiles and amphibians, and 200 species of fish. In the 1960s, it was estimated that at certain times nearly 50 percent of all migrant ducks in California were concentrated in the marsh, and that 25 percent of each year's hunting harvest was reaped here. (SSCD.1, pp. C-14, 15)[1]

The survival of Suisun is something of a miracle. In the past 150 years, California has lost well over 90 percent of its wetlands, much of it in rural areas, where population pressure is not a factor. Yet the Suisun Marsh, with Greater Metropolitan San Francisco on its doorstep, has successfully resisted wholesale encroachment. Of its 74,000 acres before modern man interfered, we still have 52,000 acres of managed marshland and 6,300 acres of unmanaged tidal marshes, nearly 80 percent of its original area. Combined with its 30,000 acres of sloughs and bays, the marsh comprises a full one eighth of all of the state's remaining wetlands. Moreover, its brackish waters provide an astonishingly rich environment for waterfowl, supporting five times the density of birds per acre as the fully salt marshes lying closer to the ocean.

This unique habitat's continued existence is due in part to some very lucky historical breaks, but in recent decades it has been the duck hunting community, especially the more-than-150 private clubs comprising all but 8,500 acres of the managed area, that have played the key role in the marsh's survival.[2]

Early Inhabitants

Before the arrival in California of the Spaniards and then the gold-seekers, the only human colony here were native American tribelets who lived year-round off the bounty of the marsh, the bay, and the surrounding hills. Suisun was their name for themselves, for

---

[1] To permit easier reading, I have abbreviated sources, which are listed in full detail in the Bibliography.

[2] Neighboring San Pablo Bay, with originally about the same acreage of bordering marshes, by the mid-1960s had been so developed that less than a quarter remained as wetlands, most of it in commercial salt evaporation ponds.

the marsh, and, many contend, for the west wind that blows here so often. (SSCD.1, p. C-3)[3] A sub-tribe of a larger group called Patwins (literally "people" in their own language), with a distinct dialect of their own, the Suisuns lived in large, dome-shaped structures that could shelter a good-sized family. (Not without some standard human problems and ingenious solutions, however: among the Patwins a bride or groom was not allowed to speak to her/his mother-in-law for the first two to three years of marriage, an excellent rule for crowded conditions and even better if—and here the record is unclear—the mother-in-law was held to equal silence.) (Johnson, p. 348)

Whatever other problems the Suisuns had, starvation was not one of them. The marsh produced not only waterfowl during the fall and winter months but fish and tule elk and edible plants all year long. The Suisuns made boats out of tules, half way between a raft and a kayak in design, up to 20 feet long and six feet wide. They were labor-intensive to build, lasted maybe a year, but, propelled by poles and sliding silently through the calm sloughs and bays, they were efficient craft for hunters and fishermen. The Suisuns used decoys for ducks and netted both fish and fowl, reserving their spears and bows and arrows for the bigger game—and, when the invaders appeared, for the Spanish troops and missionaries who had pushed their way northward out of New Spain, now Mexico, and settled in Mission Dolores.

The first white man's exploration of the Suisun Marsh was

---

[3] As this book was about to go to press, a student of Native American affairs took strong issue with the definition of Suisun as "west wind" and with some of what follows concerning 19th century relations between the Suisuns and the White Man. If my several sources were in error, my apologies. This is not a book about Native Americans, nor do I have the time or background to explore the relative merits of the conflicting versions.

conducted by Captain Juan Bautista de Anza, who arrived there on March 31, 1776, with a company of "11 soldiers, 6 muleteers, and numerous servants." He tried to trade glass beads for the large fish that the Suisuns were netting—perhaps sturgeon—but the fishermen would trade only for clothing. (Frost, pp. 9-10)

In 1810 and again in 1817, the Spaniards launched expeditions against the Suisuns. In the 1817 campaign, the natives, outgunned, retreated from the first battleground in the hills north of the Carquinez Strait to their main village of Yul Yul (not far from today's Rockville, five miles west of Fairfield), where Malica, their chief, made his last, hopeless stand againt the Spaniards' firearms. Realizing defeat was inevitable, Malica set fire to his reed hut and, singing his death song, plunged into the flames. Most of the rest of the village, including women holding their children by the hand or in their arms, followed his example. (Ibid) The surviving Suisuns then, in the way of defeated but realistic warriors throughout history, signed on as mercenaries with the victors.

Malica's successor as chief was the redoubtable Sem Yoto, a lad of 10 when captured, taken to Mission Dolores, and baptized Francisco Solano in 1810. He later turned into a six-foot-seven giant who became the dominant Native American figure in the area and whose name is now that of the county surrounding Suisun. A usually loyal lieutenant of General Mariano G. Vallejo, Solano was a vigorous mercenary who supported Vallejo against all comers, from independent tribes to uppity U.S. settlers to rival Indian mercenaries hired by the Russians at Fort Ross. In fact he turned on Vallejo only once, and, following the failure of that uprising and his own prudent disappearance for several years, returned without hard feelings to finish out his career with honor.

In recognition of Solano's later faithful service, he was issued a land grant of the area north of the Suisun Marsh, but after a few years he sold it back in 1842 for $1,000. It might have been a good investment for the Vallejo family, except that the demise of Latin rule in California was imminent.

First, in 1822, a revolution south of the border turned California residents into Mexicans, if only in name, but that lasted less than a quarter century. Then, General Vallejo's game attempts with his Suisun mercenaries to hold the territory, though temporarily successful, in the end failed. After Mexico's rejection of three purchase offers by U.S. presidents, followed by certain difficulties down at the Alamo in 1836, U.S. settlers in California "revolted" and established the Bear Flag Republic in 1846. Finally, in 1848, the older residents adopted their fourth and final citizenship by joining the United States as part of the Mexican War settlement.

For what it is worth, our government paid the Mexicans $15 million for the entire state, perhaps the book value of a couple of city blocks in Fairfield today, or, say, the whole town if one takes the devaluation of the dollar into account. Either way, a bargain.

Cities That Weren't

In 1846, Lansing W. Hastings founded Montezuma City, the intended site of a big Mormon colony just to the east of Grizzly Island. Today, at roughly this location one can find a place called Montezuma on a road-map, a small dot at the end of an abandoned railroad line. But the line has long since been torn up, leaving behind only a patchwork of decomposing cattle corrals and chutes, a forgotten siding whose rails end unevenly in star thistle, and the skeleton of a vandalized cottage, its pecked-out windows and doors staring at the desolation.

In 1847, a more serious threat to the marsh came from one Robert Semply, who drew up grandiose plans for a 400-block, deep-water port city just west of where Suisun Bay pinches into the Carquinez Strait. He named it Francisca, after Dona Francisca Vallejo, wife of the General Vallejo who deeded the land to him. So convinced was Semply that his city would be the critical link between the rich Central Valley and the ocean that he gave away his land holdings 22 miles to the southwest on some miserable, foggy, wind-swept sand dunes called Yerba Buena. Many investors from the east, convinced of Semply's cleverness, snapped up the Francisca lots, and the 300 Yerba Buenans, worried that outsiders would assume Francisca was the main settlement for San Francisco Bay, promptly changed the name of their town—to San Francisco. Still the optimist, Semply then changed Francisca's name to Benicia, which it remains today, and the town did, indeed, become the capital of California, if only for a brief 13 months in 1853-54. Thereafter, it was home to a big army arsenal and oil refineries, but the town itself has been left in relative peace as a tree-shaded backwater.

Yet another potential threat emerged in 1850, a year after the start of the Gold Rush, when a Colonel Jonathan D. Stevenson laid out his "New York of the Pacific" on the south shore of the junction of the Sacramento and San Joaquin rivers. The planning for this would-be metropolis coincided with the federal government's granting of statehood to California, a move that also explicitly put the draining, diking, and development of the Suisun Marsh into the state's hands. The state, in turn, after some years of inactivity, ceded the reclamation project to private enterprise. But by then, the New York of the Pacific had also failed to live up to expectations; not even the later downgrading of its aspirations by changing the name to Pittsburg(h) helped much, though now its tall industrial stacks do remind you of its eastern cousin.

Probably inspired by the colonel and the prospects for private development, in the 1860s a Mr. C. J. Collins laid out the 2,380-acre Collinsville about two miles across the bay to the northeast of Pittsburg and a mile southeast of old Montezuma. Mr. Samuel C. Bradshaw then bought the whole "town", changed its name to Newport City, and began energetically promoting it in New York and other major cities as the future terminus of the transcontinental railway, then a-building. He even offered to give away 500 lots free, loaning money to those who wanted to build. But Mr. Bradshaw's bubble burst (perhaps in part because many of his lots were invisible at high tide), and three weeks to the day after the Golden Spike was driven in Utah to complete the east-west rail link, the Solano sheriff sold Newport City for $29,840 to pay off mortgages. (CR) Three years later its name reverted to Collinsville, which became a salmon canning center until that industry died. (Wichel, 26 Jun 66)

For several years before World War I the slippery Mr. Patrick Calhoun, a president of the San Francisco municipal railroad who had already run afoul of the law in 1907 for bribing city officials, used his post to divert funds from the trolley system into yet another city-building scam. This time it was "Solano City," a fantasy metropolis to be built at the end of the Calhoun Cut, a three-mile dredged waterway leading west from Lindsey Slough to a point the same distance east of today's Travis Air Force Base. Unlike Newport, the Solano City site at least had the advantage of being above water, but, unfortunately for its promoter, his diversion of funds soon became public knowledge. So did his "salting" of the area with transplanted vegetables grown elsewhere to impress potential investors. Mr. Calhoun and his unfortunate produce wilted simultaneously, but his initial investment, the Calhoun Cut, remains to this day. (Coon, *West Wind,* Jul 1994)

The Collinsville and Solano City projects can be dismissed as get-rich-quick schemes that never would have succeeded anyway, but the others were serious proposals that could have spelled disaster for the Suisun Marsh.

Levees and Floods

For more than a century the marsh was also threatened with reclamation by farmers and dairymen.

In the late 1870s and 1880s, landowners began working in earnest to reclaim Grizzly Island. In those days, it was easy most of the time to keep out high water; all you needed was some coolie labor and a short stack of one-foot-thick, 18-inch square "bricks" of tule sod. After each layer of these "China" levees was laid down and allowed to dry, the workmen would add the next layer, being careful to mix the tule roots sticking out the bottom of each layer into the top of the one below to make a tight unit. Only three to five feet high, narrow at the top and steep-sided, these proto-levees in all but the most extreme combinations of tide and weather protected the grazing land that supported the early Grizzly Island dairy herds. The contract price for building a China levee was $9.00 for 132 linear feet, or $360 per mile. (SSCD.1, p. C-5)

Nowadays, to hold the floods at bay requires a levee six to eight feet high, broad enough for car travel at the top, and with no more than a one-to-three slope on the inland side to prevent undercutting when the inevitable overflows occur. As with their tule-sod forebears, those made of silt and clay are more durable than the peat levees, which tend to crack, crumble, and "squatter out" after a few years. Levees now do a better job of protection, and a modern dredge can build a mile in perhaps three weeks, but it costs more than 200 times the coolie's mile.

What made the cheap, low levees of yesteryear obsolete? After all, today's system of high-country dams, from Shasta south through the Sierra, should give us far better water control than ever existed a hundred years ago. Two factors have played the main roles: subsidence and confinement.

Old-timers say that they have seen the marsh subside about two feet in the last fifty years. In part this is doubtless due to the removal of groundwater for irrigation and other surface use. Another factor may be the very lack of flooding, which has removed a source of silt that used to build up the marsh even as it sank. Further up the delta, the subsidence is much more obvious and can measure several inches in a year. There, intensive agriculture has also played a role by loosening soil and allowing it to blow away, thus lowering the land level. Suisun, where agriculture was in slower motion, has at least escaped that fate.

Confinement has had perhaps an even greater influence on water handling problems. In the 19th century, unlike today, flood waters could spread out over hundreds of square miles of uncontrolled marshland, and it took thousands of acre-feet of water to translate into a one-inch rise in water level. Today's surface water is confined to narrowly restricted channels, where any increase in volume that can't be drained immediately has only one way to go: up. And in case anyone thinks we have put a bridle on Nature, no work of man so far has been able to handle the combination of massive Sierra snow melt, southwest winds, and torrential rains that occasionally accompany our December-January high tides. In early 1983, this unholy mix turned a scheduled 7-foot tide into a 10-footer, a memorable day for anyone who, like me, found himself looking up in horror as the bay first trickled and then swept down on him over the levee above.

But it takes more than floods to deter an ambitious farmer, as development of the rich delta bottomlands to the east of Suisun has so clearly shown. Suisun's soil, however, is not bottomland. From the outset, its salt content faced farmers with serious problems in growing crops. The hardy native grasses that were good enough for grazing dairy herds were one thing; sensitive food crops were quite another. Nevertheless, farmers eventually managed to plant orchards, wine grapes, asparagus, alfalfa, oats, barley, hay, corn, beans, melons, and berries. They also harvested the natural growth of tules and salt grass, used as packing material for pottery produced in Lincoln, across the Sacramento Valley. (Frost, p. 30-31; Huber)

Bay Damming Projects

The battle with existing and intruding salt was a hard one. No sooner would a farmer leach a field to the point it could support a crop than the bay would invade again. Discussions about shutting tidewater out of the marsh began as early as the 1860s, and in 1880 State Engineer W. W. Hall proposed a barrier across the Carquinez

Strait for that purpose. Nothing came of this, but the drought years of 1918, 1920, 1924, and 1926 resulted in high-tide salt water penetrating far upstream into the delta farmlands. These prompted a new and more ambitious proposal, first presented in 1921 by Captain C. S. Jarvis of the US Army Corps of Engineers and then revived after the 1924 drought. (Cook, p. 12; DWR 27, p. 22)

Jarvis suggested a dam from Richmond to San Quentin that would have turned both Suisun and San Pablo bays into one gigantic fresh-water lake, connected by locks to the bay's tidal parts. The advantages Jarvis and his supporters touted for the plan were creation of a permanent fresh-water supply for San Francisco, a guarantee against salt-water intrusion to Sacramento, safer navigation in the now-to-be tide-free upper reaches of the bay, elimination of saltwater borers that destroy pilings, creation of a transbay land bridge for cars and trains, and the opening of salt marshes to reclamation—for which read draining and destruction. The only negative factors they detected were the need for locks, which might have slowed ship traffic, and the possibility that a backup of fresh water behind the dam might result in levee failures in the Delta. The murder of the marsh, through drainage and filling, if anything was seen as a plus.

Again the luck of the draw, plus some sober analysis, saved Suisun. The Hetch Hetchy Dam in the Sierra was serving San Francisco's water needs, the state budget was being cut because of a recession (government spending was not fashionable as a cure for recession in those days), and anyway the drought had ended. Finally, a 1931 scientific study pointed out that any barrier to the natural flushing action of the tides in the Sacramento delta region would lead to heavy pollution in the newly confined fresh-water reservoirs. (DWR 28, p. 34)

Later in the 1930s, however, when dry conditions returned and deficit financing became popular, the Solano County Board of Supervisors launched its own fresh-water construction project, this

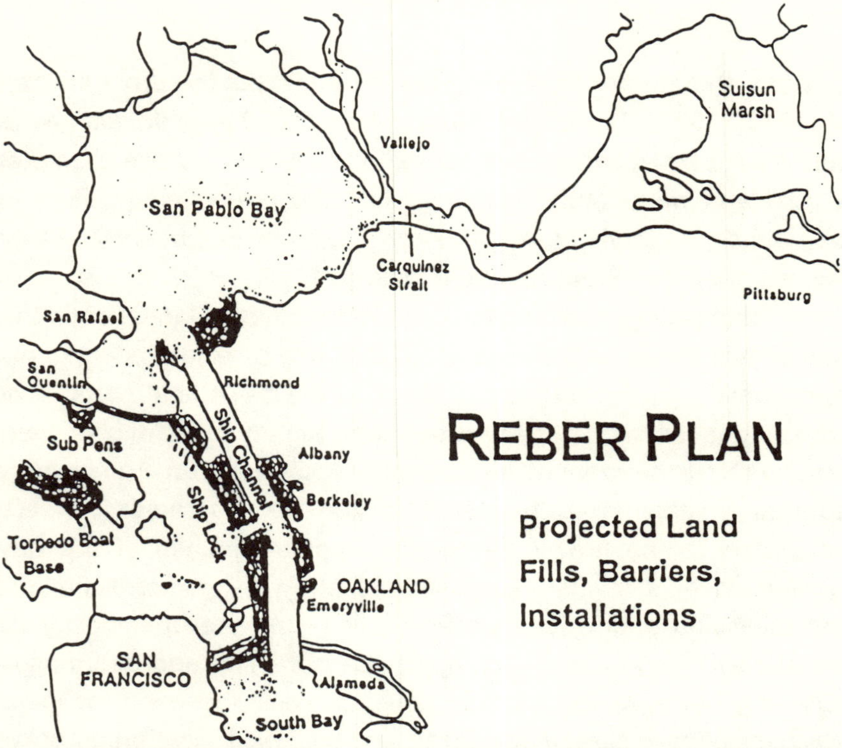

REBER PLAN

Projected Land Fills, Barriers, Installations

JOE SPINKS GRIZZLY ISLAND IRRIGATION PLAN (1936)

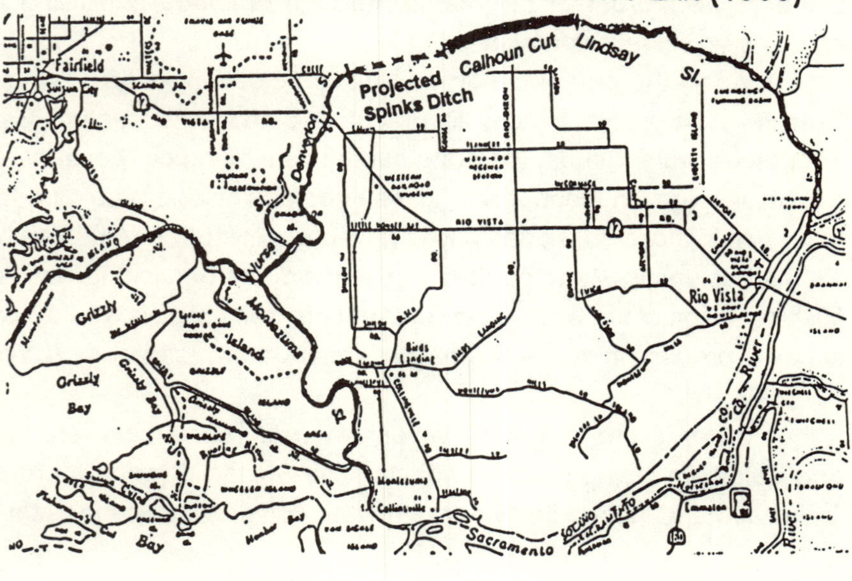

one to protect farming on Grizzly Island. The Calhoun Cut, abandoned when the Solano City project collapsed, was only a bit over two miles from another waterway, Denverton Slough, which leads south into Montezuma Slough. In about 1936 the board hired one Joe Spinks, who calculated that by damming both ends of the Montezuma Slough and digging a 2.3-mile, 30-foot-wide connecting ditch between the Calhoun Cut and Denverton Slough, he could provide continuous fresh water to the island from the upstream Sacramento. But the anticipated cost of $500,000 must have been just too high for the Solano County tax base, yet too low to fit in with the more ambitious dreams of state or federal planners. (Coon, *West Wind*, Jul 1994)

In the post-World War II years, man's tinkering with the water resource became more intense, as the state's water needs escalated. Much of the Sacramento flow, instead of helping to flush out the Suisun Marsh as before, was diverted to Central Valley irrigation or the needs of Southern California's lawns, swimming pools, and dirty cars. The marsh and the entire San Francisco Bay again suffered from the sea's invasion.

The Reber Plan

Almost inevitably, the planners sought solutions in more grandiose water management schemes. Matters reached a peak in absurdity when, in 1946, the most ambitious of all bay-damming projects, the Reber Plan, was presented to the public as the ultimate answer to virtually all the Bay Area's problems.

It was a dreamer's creation. John Reber, a retired actor and amateur theater director who had never had an hour's formal education in engineering, burst from nowhere to astound the government and public with a breathtaking plan to resculpture the entire bay region. His idea was to build not only the Jarvis-style dam from Richmond to San Quentin (the route now followed by the Richmond-San Rafael Bridge) but create another, 2000-foot-wide causeway between Oakland and San Francisco, just south of the Bay Bridge. This structure would accommodate rail lines and no fewer than four separate six- to eight-lane freeways. The two dam structures would be connected by a huge breakwater parallel to the Berkeley shore that would protect an inland waterway between the two newly created fresh-water lakes of San Pablo/Suisun and the South Bay. (Cook, pp. 14-18)

In the center of the Berkeley breakwater would be a set of mammoth locks to allow shipping into either fresh-water reservoir. The quarrying of the 20 million cubic yards of rock needed for the 50-mile perimeter of the project would provide bomb-proof caverns for military commands and hardware. In Marin County, the Greenbrae marshes would become submarine pens, and Richardson's Bay would be filled in for a torpedo-boat base. Already reduced in area drastically by landfills and drainage, San Francisco Bay would have been cut by another 85 percent.

As for his underlying philosophy, Reber had a simple answer: "This whole bay," he announced, "is too big."

Fortunately, the Army and Navy were against the plan from the outset, but the Senate Subcommittee on Public Works thought enough of it to shepherd a bill through Congress for a $2.5 million feasibility study by the Army Corps of Engineers. The *San Francisco Chronicle* backed the plan, as did the Real Estate Association of San Francisco, various civic groups, Delta farmers, several military officers, a dozen recognized Bay Area engineers, and some Suisun landowners concerned with the saltwater intrusion that was again degrading their soil.

Although initial enthusiasm for the Reber Plan was high, it began to fade as various studies pointed to serious problems: the predictable annual evaporation of a million acre-feet of water in the new lakes, the end of circulation in the South Bay and its conversion into a giant cesspool, the silting up of San Pablo Bay, the closure of the Golden Gate as a huge, impassable sandbar built up offshore, and the kill-off of migratory fish. (Regarding the last, Reber sniffed that the new lakes could be replanted with every fresh-water species and become "without doubt the greatest fishing hole in the world.")

In the end, the project was killed, but two lasting spin-offs remain to this day: the elaborate large-scale model of the bay and its circulation system, bought with the federal study funds and housed in Sausalito; and the perception that *some* kind of long-range planning for the bay was necessary. Ironically, the second of these resulted in the formation of the Bay Conservation and Development Commission (BCDC), which since its creation in 1965 has focused most of its energies on preventing runaway development and the environmental pollution that comes with it.

It is unlikely that Mr. Reber, who died in 1960, would have approved of BCDC—or even understood what all the fuss was about.

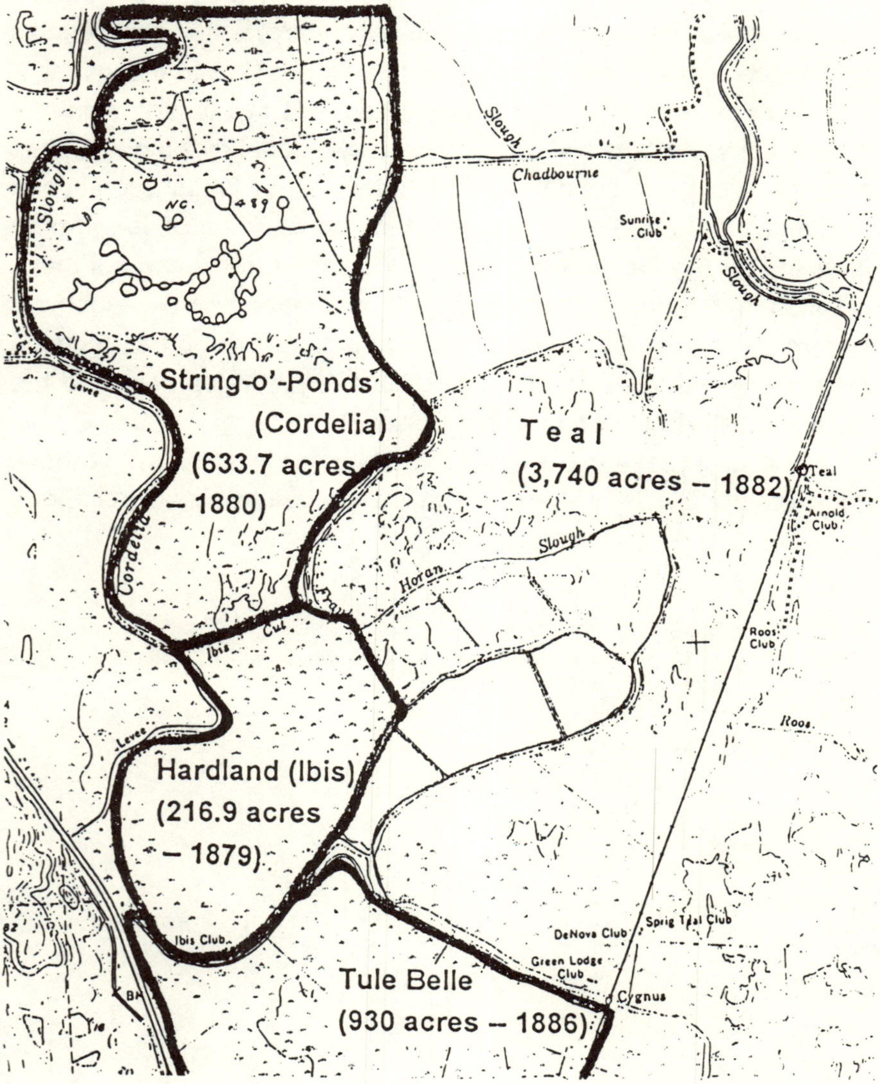

THE FIRST (PRE-1890) SUISUN MARSH CLUBS

(Teal and Tule Belle extended beyond this map's borders)

CHAPTER 2

EARLY HUNTERS, RAILROADS, AND CLUBS (1879-1892)

The Indians had no firearms, so the first man to *shoot* a duck in the Suisun Marsh was probably a Spaniard. (Or it might even have been Chief Solano, who in his declining years retired to his own adobe house at the marsh's edge.) In any case, our record only goes back to the market hunters who began supplying San Francisco with Suisun ducks in about 1859, and who monopolized the shooting grounds for the next twenty years. Four of these deserve special attention, Jim Payne and Seth Beckwith for having leased the best west-side marshes that later became sites for the first duck clubs, William W. Richards for becoming the most consistently communicative Suisun old-timer, and Walter Welch for giving us the best details on market hunting and early wardening.

Market Hunters and Day Shooters

For the market hunters, it was not a sport but a business, and as Richards put it, "in those days it was not a matter of how many ducks we could kill, it was how many we could get out to market." This was no easy problem. Welch recalled that in 1879, when he first hunted in the Delta, it took his slow-moving scow a solid week to make the round trip from San Francisco. Suisun was only two thirds as far from the City, but it was a day's travel either way under the best conditions, and more when the weather kicked up.

Welch's account of his market hunting is unique and must serve as our model for his Suisun colleagues, who left no known records. He and his partner had six vessels: a 42-foot, four-berth "scow sloop" with an 11-foot beam for a headquarters, two tule splitters (narrow, double-bowed boats, also called pickaxes, for polling through flooded grasses), two scull boats, and a small, fast sailboat for hauling their take to market in San Francisco. They brought along 300 decoys, a ninety-day supply of staples, and a whole arsenal of guns and ammunition: one 4-gauge and one 6-gauge muzzle-loader, two 10-gauge and two 12-gauge breech-loaders, two Winchester 45-70 rifles, 100 pounds of powder, 500 pounds of shot, 10,000 shotgun wads, 5,000 primers, and 500 big brass shotgun shells.[1] (Welch, 1927, pp. 180-184; 1931, p. 256-259)

The supplies were none too many. The 10-gauges, which were their main weapons, used 1 1/2 ounces of shot and 5 drams of powder a pop, a load that once netted Welch 38 teal with one barrel. The double-barrel 4-gauge, reserved for big flocks, consumed 20 drams of powder and 6 ounces of lead for each shot. The 12s were just used for finishing off cripples.

They must have been iron men, those market hunters: the 4- and 6-gauge were hand-held weapons, and the 4-gauge's record for one barrel was 89 "mixed ducks" and for both barrels 120. Welch does not mention the kick, which was probably enough to send

---

[1] Those reusable, full-length brass shells are still common in Russia and Central Asia, where they are customarily overloaded with black powder and whatever old metal comes to hand for the shot; when touched off they produce a 15-foot firework of sparkling flame and a clap of thunder.

today's average hunter back to about last Wednesday. Under good conditions he and his partner normally killed 100 to 200 ducks apiece per day; if they had not shot 200 to 250 shells between dawn and 1 PM, the hunting was below average.

For Welch, the season in the delta ran from September 1 through November. The partners then hauled anchor and came back to San Francisco Bay for the diving ducks until the first of March.

Prices for game on the San Francisco market varied according to the species: $.75 to $3.00 a dozen for ducks depending on the species and size (teal at the bottom of the list, canvasbacks and sprig at the top); $1.00 to $4.50 a dozen for geese; swans and sandhill cranes up to $6.00 a dozen; $.35 to $.75 a dozen for shore birds, bitterns, blue herons, night herons, and all the smaller waders (strange tastes our ancestors had). (Ibid)

By the time Welch came along, Payne, Beckwith, and others had already cornered the best market hunting close to San Francisco. This was an expanse of wetlands along the west side of the Suisun Marsh, between Suisun and Cordelia Sloughs.

In 1871 the Berkeley-based Chamberlain family got a U.S. patent on the property, thereafter leasing hunting rights to the same hunters who had been using it for nothing. Before the end of the decade, however, the senior Chamberlains had died, leaving the land to their three minor children, Mary, Emily, and Freddy. In the relaxed fashion of those times, there was no official registration of the hunting lease, but the market hunters still had to pay the Chamberlain estate (with understandable anger, one imagines) for the privilege of harvesting the ducks.

Embryonic Clubs

Possibly in part to offset the leases, Payne in the 1870s began building up a limited clientele of rich San Franciscans to guide as sports shooters. Before, most San Francisco amateurs had hunted the open marshes on the east side, where the Sacramento and San

Joaquin emptied into Suisun Bay, but sedimentation from Sierra mining and valley farming was already beginning to drive waterfowl away from these traditional grounds. Payne charged $5.00 a day, no minor fee at a time when a single dollar could buy a man's hard labor from sunup to sundown. But for San Francisco's businessmen the cash was negligible compared to the cost in lost office time of a tedious mule-cart or boat commute at either end of every hunt.

Nevertheless, many were willing to make that sacrifice. An interesting sign of the times was how the best shooting ponds began changing names. As a keen teen-ager, M. Hall McAllister—later one of California's great hunter-naturalists—in the late 1870s mapped some of the ponds in market hunters' names (Judd, Hayward, Smith, and Montgomery), while according to Richards they were already being renamed for such prominent San Francisco amateurs as State Senator W. W. Traylor, stockbroker Albert T. Titcomb, and Commander Richard S. Floyd, late of the Confederate Navy and in California the constructor of the Lick Astronomical Observatory atop Mount Hamilton. (Richards, 1931)

By that time, Richards, an upwardly mobile entrepreneur who had an early 900-acre lease from the Chamberlains, was already associating with the rich and famous of San Francisco. McAllister, by contrast, was being raised in more modest style by his uncle, a colonel at the Benicia arsenal, and he stuck to the old names well into the 1880s.

Other ponds quickly came to reflect San Francisco high society (Wilson, Whittier, Hopkins, Josselyn, Gerber, and Sinclair), yet others were named for locals (Thickbroom, Boynton, Morrill, Wing, and Pringle Ponds, to cite just a few), and the Sixth Reach Pond, termed "the largest closed body of water in Solano County," got its name from the sixth tack a sailing vessel had to make in navigating the Suisun Slough. Many other ponds drew their names (Hardland, Lower Basin, Hospital, Island, Northeast, Lower Surveyor, and Upper Surveyor) from unknown sources. (Ibid; Stoner, p. 245)

McAllister's hunting log provides a wonderful feeling for his life as a young hunter. In 1877, when he was just turning 16, he carefully recorded each "pass" he had received to leave the arsenal for a day's hunt, his allowed absences running from 16 hours (2AM to 8PM) on a generous day to a short 11 hours when somebody felt it was necessary to rein him in. The arsenal was only five or six miles from the marshes, but the trip in those days was by mule cart, which could take as long as three hours each way.

At first, his take was broad, undistinguished, and not always identifiable. For example, on 22 September 1877, his first day out, he bagged 4 "peeps" (killdeer?) in addition to 2 snipe and 3 plover. Later, he and his companions (he usually shot with one to three others) began to get more ducks, then a few geese, and before long he was hitting more than his share. But the bags were low, averaging only about four birds per gun for his first season.[2] Sometimes the weather forced him to seek shelter in Payne's "ark" (houseboat), which was anchored near what would become Teal Station, and on one occasion Payne helped him push a boat across the bottomless mud of Sixth Reach Pond. For the most part, however, he hunted Jabez Thickbroom's marshes (today's Tule Belle Club), which could be reached without crossing Cordelia Slough. (McAllister scrapbook)

On 22 October 1879, his 18th birthday, McAllister moved to San Francisco to live with other relatives, the Hall McAllisters. This threatened to put the Suisun marshes beyond his reach, but before the next season, direct rail access to the marsh made quick round trips again possible, not only for him but for some richer rivals. The construction of the Central Pacific (later Southern Pacific) rail link

---

[2] McAllister's poor bags in 1877 were just a sign of a young boy starting out. That same year Joe and Henry Bassford reportedly got 184 cans in one morning at the nearby Hardland Ponds. (*B & S*, 12 Feb 1887)

marked a crucial turning point in marsh history.

The Railroad's Role

By boat and mule cart, hunters had been getting to the Suisun Marsh for years, but the railroad brought its west side to within a short commute from city offices. In 1879 the transcontinental rail link was just ten years old, and the railroad moguls decided to shorten and speed the western end by cutting out Vallejo.

From Fairfield, the existing line already went southwest through Cordelia. From there, a direct link to Benicia (the route taken by today's Interstate 680) would have provided an ideal surface of flat bedrock along the foot of the Coast Range for the new roadbed. Instead, the engineers laid their tracks directly from Benicia to Fairfield, a slightly more direct route but one that traversed one of the most railroad-unfriendly environments south of the arctic tundra.

Construction began early in 1878 and was supposed to take less than a year. Late in October 1879, with six miles of the new track sinking steadily into the marsh, the gravel train's operator, "a jolly fellow who has just gotten acquainted with the (local) boys and possibly with the girls," was pleased to learn that his equipment would be needed for at least four more weeks. (*SR*, 23 Oct 1879) The date of the actual opening of the line is not a matter of known public record.

Delays in completion were a minor matter compared to the troubles that followed. For the next 35 years the tracks sank a foot a year as piles set on piles to support the trestles vanished without trace into the mud. For three months in 1905 the line was shut down for massive reconstruction and filling. The day before it was due to reopen, a thousand feet of track disappeared from sight. The next year, the San Francisco earthquake dropped one section 12 feet. Endless infusions of rock and a million board feet of lumber failed to hold the line. In late 1911, a hundred feet of track at the well-named Joice Sink was ten feet under water; a few weeks later, a heroic road gang working around the clock elevated the roadbed from four to 12 feet at Teal and Cygnus Stations—only to find the following dawn that overnight subsidence had again drowned their work. Duck hunters marooned north of there had to be evacuated by trains that ran backwards from Suisun and eventually brought them home the long way via Vallejo. (Wichels, 1966, 1974; *B&S,* 18 Nov 1911)

From May 1913 until January 1915, the company replaced most of the trestles and cribbing with uncounted tons of fill to spread the roadbed seven feet beyond the ends of the ties on each side. This maneuver allowed the tracks to "float," (they even rose and fell slightly with the tide), but it achieved only relative stability. At last, in 1925, the railroad claimed its war with the marsh was won, but in the late 1980s a maintenance official told Teal Club keeper Sam Hartzell that upkeep on the Benicia-Suisun stretch of track was still the most costly operation for its length in the whole United States.

Why did they do it? The moguls who owned and ran the Southern Pacific—Huntington, Stanford, Hopkins, and Crocker—were known as ruthless tycoons whom one later historian, Lucius Beebe, likened to medieval barons in Italy. Their castles on San Francisco's hills, he wrote, symbolized their absolute hold over all who lived below them. But their foolish commitment to running their trains across the marsh may reflect the influence of San Francisco's uppercrust duck hunters, who had good reason to tell them how—or at least where—to run their railroad. The moguls themselves were not hunters, or they surely would have jumped at the Chamberlain estate's 1875 offer to sell them its entire marsh holding for $.75 an acre. (McAllister, 1932)

Before the tracks were even laid, the railroad agreed to open two stations, Teal and Drawbridge, and Payne and Beckwith had anchored their sloop *Wave* at the site of the former. (McAllister 1930) As late as 1882, Teal was still the only stop, but Drawbridge (later renamed Oluta, then Cygnus), Goodyear, and Pintail (renamed Jacksnipe) eventually followed. These stations' only function was to provide access to the "hot" duck hunting ponds pioneered by the market hunters, none of them farther than a half mile away from the line. While Teal was the only stop, the newly established clubs arranged with the railroad authorities to reserve it for themselves. "All the ponds above and below the tracks are preserved for private parties. The best shooting places near the city for those desiring a day or two's shoot... are now closed to the public," noted the *Breeder and Sportsman* sadly. (*B & S,* 8 Jul 1882, 15 Jul 1882) The gentleman hunters apparently had the power not only to set the course of the transcontinental tracks but to guarantee privacy by monopolizing their stations. (*B & S,* 8 Jul 1882)

They then set about establishing the unique hunting societies that sprang up beside the rails, societies that might never have developed without the handy—almost door-to-door—commute from the city.[3]

The First Duck Clubs

Even before the first roadbed had been completed and begun to sink, the first group of private sportsmen moved in. Once the line was open, other groups followed. Like blackbird flocks in a newly flooded marsh, they swirled and mingled and separated in a feeding-frenzy confusion of shifting alliances and territories.

As the "gentlemen" moved in, the doomed market hunters quickly became extinct. Unlike the rich amateurs, they were not to be found in the *Blue Book* or *Social Register*, nor, within two years of the railroad's coming, were most of them found on the marsh, either. Of

---

[3] In 1896, deliveries of wild ducks of all species to the San Francisco market were two and three times greater from Yolo and Merced counties than from Solano, but their marshes, unlike Suisun, had only indirect rail service. Market hunters were willing to lug their kill some distance to a station, but private clubs came in only after access improved. (*B & S,* 14 Nov 1896)

them all, only Bill Richards survived as a hunter. Except for a New York interlude from 1890 to 1900, he remained a fixture on the marsh from the time he first shot it in 1869 until his death in 1936, nearly 70 years later.

The first club to form was the Hardland (a misnomer for *any* piece of the Suisun Marsh) in July 1879, before the railroad was even completed. Its charter members included Richards, and its clubhouse was on Cordelia Slough, about a half mile west of where the rail line crossed. By September 1880, Payne had formally subleased the property to six of the Hardland founders and a newcomer, William B. Bradford. (Stoner, 1937)

In January 1881 the Chamberlain heirs leased the same property to a Henry C. McPike, who then hastily transferred the lease to Charles Toland in April 1881 and was heard of no more. By the 1882 season, Toland and two other newcomers had rechristened it the Ibis, as it is still known. The name derived from the white-faced ibis, a bird then common in the area, and from the members' charm with something that hinted of old Egyptian sacred rites.[4] (CR, Stoner)

Curiously, despite the first years' shifting ownerships and name change, the Ibis is the only first-generation club on the marsh whose territory has remained virtually unchanged throughout its entire history. By the early 1900s it had a flag (a white ibis, rampant, on a red field) a cartoon coat-of-arms (a dancing hunter with a duck as a partner), a pass word (MARK!!!), and a motto ("Let 'em light" [tsk tsk]). (McAllister, *scrapbook*)

Even though it underwent a re-christening, the Ibis thus challenges the Cordelia Gun Club's claim to being the oldest club on the marsh. Nevertheless, the Cordelia unquestionably has the oldest

---

[4] Stoner's claim that the Ibis was originally the Canvasback is wrong. Several sources have identified the Canvasback as the precursor to the Family Club. (*SR*, 1 Dec 1899; *B&S*, 17 Jun 1905)

Photo courtesy Suisun Resource Conservation District, Identifications from M. Hall McAllister hunting scrapbook.

CORDELIA CLUB FOUNDERS

Clockwise, starting at center top: Edward F. Bent, grain merchant and club president; M. Hall McAllister, import-export merchant; Fred S. Butler, insurance agent; Wm. B. Bradford, Alaska Packers Assn; John K. Orr, "sportsman and haberdasher;" Charles W. Kellogg, club founder and manager of Tubbs Cordage (ship rigging) Co.; Harry Babcock, Babcock Estate; Ward S. McAllister, attorney; T. Carly Friedlander, "son of Isaac Friedlander, California Grain King;" and Wm. McPherson, grain broker.

name, having held it since the club's founding in 1880 in San Francisco's Occidental Hotel at a meeting chaired by Charles W. Kellogg. Ten hunters attended, including two Hardland founders and William Bradford, who clearly did not believe that one duck club was enough. (A bachelor, no doubt.) (McAllister, 1930)

But the Cordelia cannot have been where it is located today. The present 640-acre property, described as containing a "string of ponds," was first rented from the Chamberlain estate by Bill Richards in June 1881. There, he set up the String of Ponds Club with six of the Hardland founders, who by then had been replaced at the Hardland by Toland and his friends. Richards subleased this marsh to Cordelia's president, Ed Bent, only in March 1882. (CR)

Until then the Cordelia shared shooting grounds with the Teal Club. Instead of fixed clubhouses, the clubs used floating quarters. The Cordelia leased Captain Charles Chittenden's yacht, the San Francisco-based yawl *Lolita*, which they moored near Teal Station, next to Jim Payne's houseboat and sloop *Wave*. On the *Lolita's* deck they perched a shack for their accommodations during the hunting season. This ungainly superstructure was not permanent, however; every spring it would be dismantled and the boat would hurry back in time to open the racing season in San Francisco, where it was a respected contender.

The Cordelia kept its housing near Teal until 1888. After the first three hunting seasons, the *Lolita* was replaced in 1883 by another vessel, *White Wings*, and still later by "a large and commodious house ark." In any case, from 1882, when Bent acquired the lease, the Cordelia shooting grounds were where they are today, but the floating clubhouse remained at first near Teal and then anchored further away, probably closer to the shooting, on Frank Horan Slough. The club's present clubhouse on Cordelia Slough became their headquarters in 1908. (McAllister, 1930)

In one sense the Teal Club can also lay claim to being oldest, though it was formally established only in 1882. Before the railroad was built, Payne and Beckwith for some time had moored their sloop *Wave,* anchored their "ark," and stored their gear where the railroad had promised it would put in the first station, Teal. This was close by the "Payne ponds" that their San Francisco guest hunters had already begun to rename for themselves: (G. Frank) Smith, (William) Gerber, Floyd, Titcomb, Whittier, Taylor, Traylor. There is little doubt that the paying guests, who regularly overnighted at the ark, felt proprietary toward "their" ponds, and long before they banded together formally under the chairmanship of G. Frank Smith, they must have shared that sense of camaraderie which is the essential base of any club.

When they finally overcame Payne's reluctance to part with his lease, the eight founders of the "Teal Shooting Club" were merely continuing an existing association. On 22 July 1882, they bought the Payne lease and all the market hunter's equipment for $2,000, immediately sank another $5,000 in still more arks, boats, and unexplained "privileges," and on 30 August, the club was formally founded. Jim Payne was retained at a salary of $100 a month to manage the property.[5] (*B & S,* 30 Sep 1882)

From the outset, the Teal went first class. Payne oversaw a staff of four, including two men to clean members' guns, row them to their blinds, and do chores, a Chinese cook, and a Japanese steward. Each of the eight members had his own stateroom in the new ark built for the purpose, with "names on the doors, with washstands, hair mattresses, brussels carpets, etc" (Ibid)

The first club known to own, not merely lease, its marsh was the Tule Belle, just to the south of the Chamberlain tract. The original

---

[5] One of the arks, whether Payne's or a "new" one, still underlies part of today's Teal Club sleeping quarters. The four-inch, round-head, squared-off nails that still hold its last redwood planks in place are rusted on the outside, but where they are still embedded in sound wood, they are as shiny as when they were first pounded in.

property owner was Andrew Goodyear, who obtained it as a land grant in 1880. He rented and later sold part of it to Jabez and Agnes Thickbroom, who gave a "long lease" to Tule Belle members Ellis H. Holmes and Oliver N. Bogart in 1885. The club originally had been sited in and named for the town of Belmont near San Mateo, and then, in about 1880, moved to Sherman Island where it took on the name of its barge-clubhouse. But the shooting on Sherman had deteriorated so badly due to sedimentation from the Sierra mines that the owners moved their club—barge, name, and all—to the new property in February 1885, and offered up two new annual memberships for $250.[6]

By 1886, however, Thickbroom had fallen on hard times, and the property reverted to Goodyear, who sold it to Holmes in May 1886. Holmes, in 1889, sold the property to his fellow member Charles Josselyn, who expanded its acreage from 400 to 1,000 after Thickbroom defaulted on another mortgage for his remaining marsh property in the mid-90s. (*B & S,* 24 Nov 1883, 6 Mar 1886; CR; Richards, 1932)

In short, the oldest known fixed duck club property is the Ibis (1879), the oldest name is the Cordelia (1880), probably the oldest hunters' association and certainly the oldest clubhouse still in use is the Teal (late 1870s or early 1880s), and the oldest club to own its marsh property is the Tule Belle (1886). Along with Richards' String of Ponds Club, they were the seminal duck hunting organizations on the Suisun Marsh.

Developing Tensions

The club members lived high and they died high. None went

---

[6] See Chapter 7. The sediment from mining was estimated in 1957 to have totaled one billion cubic yards of sand and silt, which precipitated in Suisun and San Pablo bays up to depths of three feet. (Gilliam, p. 85)

out in grander style than did Teal's Commander Floyd, who expired in 1890 in Philadelphia when he was only 47. They shipped him back to San Francisco in a casket so huge that the Pullman car reserved for him and his surviving widow had to have a side panel sawn out to get it aboard. (The casket was, mused the *Call* with quiet civic pride, "one of the finest ever seen in this State.") But deaths—then as now—could not keep up with births and immigration, and as more people arrived in California, hunting pressures inevitably rose.

For a while, property rights on the marshes did not count for much, as witness McAllister's free ranging and the sparse frequency and sketchy conditions of the few formal leases that were recorded. But from the mid-1880s, as the urban rich from the coastal counties leased more and more Suisun marshland for their own, they began to crowd out competitors. Leases became more formal and were registered. "No Trespassing" signs went up. A species of intra- and inter-class warfare broke out, a three-way battle for the marsh that pitted the new clubs against local hunters, against each other, and against other city hunters.

At the end of the 1892 season, the *San Francisco Examiner* had its own version of the worst that could befall a hunter:

> The improbably long shot and phenomenally heavy bag are being discussed at every city club, but no man tells the grisly tale of finding, when he arrived where the ducks were thick as blackbirds in a barnyard, that he had brought 98 ten-gauge shells into the field and had an 8-gauge gun. There are moments in a man's life that he does not wish to recall. (*SFX,* 1 Mar 1892)

By the end of the next season, hunters' worries had become considerably more serious.

Chapter 3

War and Peace on the Marsh (1883-1899)

The complex of hostilities that developed on the marsh during the 1880s and 1890s would have made a Balkan warlord blush. It was a small, mostly bloodless war, and it was based, like so many bigger wars, on greed for a limited resource and envy of those controlling it. The resource was nothing less than the best duck hunting land close to San Francisco. Men have died for much less worthy causes.

At the root of the problem was the whole question of whether duck clubs ("preserves") should exist. As early as 1883, the San Francisco weekly *Breeder and Sportsman* carried a thoughtful article outlining the contradiction between the public's claimed right to hunt on open, unimproved marshland (later, for a time, to be confirmed in law) and the exclusive rights claimed by the titular owners of that land. The article reported that "growls" against the preserve system were already audible locally, yet it had been found on the already crowded East Coast that preserves were the only way of saving wild game from annihilation from overhunting. It recommended following a practice developed in Wisconsin, where rich landowners issued $5 seasonal permits to responsible hunters, the fee being designed not to reap a profit but to give the owner some control over the number and character of the hunters allowed on his land. In its way, this suggestion forecast the modern state-run licensing and special-permit-issuing offices. (*B & S,* 10 Nov 1883)

The raw frontier democracy of the West in those days did not look kindly on such reasonable solutions, however, and the growls soon escalated into legal shoving matches. A serious falling out between the biggest and most influential Suisun preserves, the Teal and Cordelia clubs, made things easier for grassroots rebels.

Origins of the Teal-Cordelia War

Not all the early leases were matters of official record, but there is little doubt that the Teal, which got in on the ground floor and had most of the acreage from the start, paid far less per acre than the others. In late 1884, the rumor floated that the cheap, $125-per-year Teal lease was to expire in July 1885 and that a bidding war for the property with other "gentlemen" (unidentified) was about to begin. (*SF Call,* 29 Dec 1884)

With considerable truculence, Teal president G. Frank Smith responded that he already had a signed, legally airtight lease that would begin on 22 July 1885, and that the club would occupy its premises *for the next nine years* (his emphasis), "peaceably if we can or forcably (sic) if we must." Any effort to divest the club of its lease, he said, would be "wasted coin."[1] Sure enough, when the 1885

---

[1] One gets the impression that Smith was not a very likeable man. Of the eight Teal founders, one died and three resigned within the first two years. Payne lasted less than one season as manager before going off to found another club on the Petaluma marshes. A news item dated 15 December 1888 gleefully records how the "ever-modest" Smith, after boasting of being the Teal's champion shot, was bested by fellow Teal member Edwin Goodall. (McAllister scrapbook. Newspaper not identified.)

leases for the Teal, Cordelia, and Ibis were all signed at once on 22 July, the Teal, even though its rent had doubled, again emerged with the best bargain, which must have rankled the Ibis-ites and especially the Cordelians. (*B & S*, 3 Jan 1885; CR)

Nevertheless, the floating clubhouses of the Cordelia and Teal temporarily continued to be moored next to each other, and there is no evidence of immediate hostility between the two. Later, however, there was a bitter dispute over shooting rights to a particularly productive pond, and in 1888 the Teal peremptorily demanded the Cordelia move its sleeping quarters away from the Teal's moorings. This meant that the Cordelians no longer had the easy, 15-minute row from Teal Station to the place where both had tied up together, probably the junction of Teal and Frank Horan Sloughs.[2] (*SFX,* 8 Sep 1893)

The Local Partisans

Meanwhile, all of the absentee-owner clubs were suffering from the depredations of local poachers. From the time they were founded in the 1880s, the Ibis, Teal, Cordelia, and Tule Belle clubs had each had to retain two to three keepers, post their territory, and run off intruders. In 1883, the club-owners' cause was taken up by the *Breeder and Sportsman,* which blasted the "maraudings of thieving poachers... (who) devastate (the clubs) for five days and nights every week so that ... at the end of the week there is not a bird to be found." (*B & S,* 22 Dec 1883)

---

[2] Possibly associated with the feud are two peculiar entries in the Solano County Book of Assignments, showing that Kellogg passed the Cordelia lease to an Englishman, John Lee, in 1886, and Lee passed it to another Englishman, J. J. Palmer, in 1888, conceivably to avoid legal responsibility for the club. There is no record of the lease's later reversion to an American. (CR)

Courtesy, San Francisco Maritime NHP, Classified Photograph Collection, Cl. 14,150n.

Lolita (left) and *Wave* peaceably at anchor together in 1881 near Teal Station. Season is on (note swan hanging from *Lolita*'s bowsprit), but there is only an awning for shelter. Later, a functional but hideous clubhouse/shack was erected on her deck each hunting season. The man in *Wave* may be Jim Payne. The "ark" in the background is probably the present foundation for the Teal Club's sleeping quarters.

But neither criticism by the magazine nor cursing by the keepers had any effect, and stronger measures were adopted. "Remonstrance seems ineffectual. One of the keepers on one of the preserves, on hearing shooting at night, went down and fired ten to twelve shots with a rifle across the ponds. On the next occasion when he went, a dozen or so shots were fired at him from a revolver, which he answered with his rifle. The bullets flew around quite lively, but no one was hurt, it being too dark to see persons. Since then, however, the preserves in question have been undisturbed." (*Benicia Era*, as quoted in *B & S,* 6 Dec 1884)

By September 1885, *Breeder and Sportsman* was openly advocating that the clubs "station a keeper on some commanding elevation, and instruct him to practice with his Winchester on anything resembling a man or a duck boat on days when members were not using the ponds." When keepers began doing just that—and being pilloried for it in the rural press—the magazine strongly defended them. In late November, poachers on the Cordelia club were met with bullets but responded by returning to the marsh with their own rifles, which deterred any further gunfire from the keepers. Miraculously, there were no reported casualties in any of these exchanges. The hostilities probably reached these extremes because the mid-1880s seasons were very poor ("Where have the ducks gone?" bleated the press, in a refrain still occasionally heard around the marsh). Though warm weather and wet storms doubtless contributed to the bad shooting, the poachers caught most of the blame for frightening the ducks.[3] (*B & S,* 17 Jan, 19 Sep, 7 Nov, 28

---

[3] Other local sources of annoyance to clubs were hogs that were turned loose on the marsh in the spring and summer, where they ate up eggs and ducklings, and arsonists who set fire to tules. By late 1886, the clubs had banished the hogs and somehow deterred the burners. (*B & S,* 27 Nov 1886)

Nov 1885)

Eventually, in late 1892, the locals decided to form their own clubs. The first was the Suisun Gun Club, "composed of energetic young men who have placed their organization on a sound business foundation... (and who have) rented the Tomasini tules (and) erected a club house. They propose to have a formal opening of the club house in the near future." (*SR*, 2 Dec 1892) But after that one brief announcement, nothing more was heard of this organization until it was reincarnated in 1896 as a new club with the same name. The project probably hung fire because the locals thought they might profit from new legislation that would allow them to shoot the rich men's preserves with impunity. In this, they were counting on some help from within the enemy's camp—other city hunters who also wanted to abolish preserves.

The Urban Enemy

Hunters from the city could be divided into two groups. The first were what today are sometimes termed limousine liberals, reformers of wealth and status but ideologically to the left of those other, more conservative rich men who held a monopoly over the Suisun shooting grounds. The reformers relied on organization, public relations, and the courts to promote their cause.

But the reformers also included in their ranks those who today might be termed rednecks—boisterous, poorer individuals who demanded their rights more directly and who had few qualms about resorting to violence when thwarted. They were later lumped together in the press as "unattached." There was no clearcut dividing line between the liberals and the rednecks, who blurred together in the center, but neither could have been too comfortable with the methods of the other when conflicts with the baronial club owners heated up.

Most vehement of the limousine group in condemnation of private clubs was the Sportsmen's Association of California (SAC), which was founded in San Francisco on 15 February, 1889.[4] The preamble to the SAC constitution and by-laws stated that the association would be devoted to resisting further encroachment by "selfish individuals who (are) annexing all available tidewater land to establish private game preserves." SAC felt that this was "foreign to our country and inimical to the interests of true sportsmanship." Among other complaints, SAC accused the clubs of having their keepers slaughter ducks for the market. (*B & S,* 23 Feb 1889; SAC Constitution and By-laws, 1891)

SAC asserted that it would operate within the law and that its goals also included strict enforcement of existing game laws, protection of the state's native fish and game, and introduction of new species as appropriate. Anticipating some problems, the association and its president guaranteed to defend "any member... in good standing, arrested for trespassing on illegally reclaimed and uninclosed land." (Ibid)

Nowhere are any of the accursed private preserves named, but the clubs that had already appeared on the Suisun Marsh certainly fit the association's definition of its enemies, and before long they became SAC targets. It was, perhaps, more than merely a matter of principle—the huge bags the Teal, Cordelia, and Ibis had been reporting were probably just too tempting to pass up—but SAC wrapped itself in a banner of righteousness as it promoted its aims.

SAC could capitalize on local support in its war against the old clubs. The *Vacaville Reporter,* for example, blamed the absence of ducks in 1892 on "San Francisco capitalists who call themselves

---

[4] The first of our state's sportsmen's organizations was probably the conservation-oriented California State Sportsmen's Association, formed in 1881. There followed the Sportsmen's Protective Association and then SAC. All promoted shooting for the common man, but SAC was the most outspokenly anti-club.

hunters" and who had wiped out the birds. "I am told that some of these fellows... actually hire the keepers to go out the night before and put out fish lines (sic). When morning comes they go to the pond where they find ducks fast to the lines, whereupon they kill them with their expensive guns and then haul in their lines. The state ought to set aside part of this marsh land for (public) hunting purposes...before it is all gobbled up by these 'dude hunters'" (*VR,* 28 Jan 1893)

The reaction of the conservative San Francisco *Breeder and Sportsman* to the gun club controversy was moderate. Referring to agitation to abolish them as "the same old decidedly nutty flurry," the magazine strongly defended the preserve system, but it also said that SAC counted among its members many men who were "scrupulously honorable in business and social relations." Given the editorial practice in those days of applying black or white paint with the broadest of brushes, it was a surprising concession. But the bottom line was the defense of property rights, which it justified in part by noting that the club owners were "sportsmen of the best type, (who) own fine guns, wing shoot exclusively, (and) are generous in extending kindnesses to non-members." Besides, it added, they paid heavily for their sport, and usurping their territory would lead to anarchy. (*B & S,* 23 Feb 1889)

SAC might have died fairly young if it had not been for the Emeric Bill, which passed the Sacramento legislature in March 1893. It was a sweeping game law that outlawed the taking, possession, and sale of all game in the off season (1 March to 1 September), the use of any means but a gun not larger than 10 gauge to take game during the open season, and hunting on private and posted "enclosed cultivated grounds" without permission. The hooker was in the last clause, which implicitly opened the entire Suisun Marsh, which was neither fenced nor cultivated, to public hunting. (*B & S,* 18 Mar 1893)

And just a few months later, festering bad relations between the Teal and Cordelia clubs erupted into an open feud that suited both SAC's and the local Suisun hunters' purposes admirably.

Open Teal-Cordelia Warfare

When the Teal lease came up for renewal on 22 July 1893, the Chamberlains—possibly at the Cordelia's instigation—tried to raise the rent from $250 a year to $1,200. This was during the worst economic depression thus far in U.S. history, and the Teal, either strapped for cash or perhaps assuming the Chamberlains were bluffing, refused to ante up. As the lease ran out, the Cordelia Club was there, checkbook in hand, to snatch away the entire Teal holding. They got more than they had bargained for. (*SFX,* 8 Sep 1893)

The shocked Teal, not about to play doormat, recovered swiftly with a massive counterattack. First they arranged with the Southern Pacific to lease 1,200 feet of right-of-way near Teal Station (at that time possibly still the only station on the marsh) and quickly moved all of their buildings onto their new property. They then fenced this off and installed a low bridge and floating dock to block any boat traffic up Teal Slough, the only quick way one might get from Teal Station to the Cordelia shooting grounds by water. (Ibid) Next, they made instant use of the Emeric Bill, especially the opening of the duck season on 1 September (the only such early opening since 1853 and never to happen again). (CA Code, 1893)

Also, in a paper move to avoid any Cordelia lawsuit, they formally disbanded as a club, turning over all their buildings and equipment to their "manager," one James Brundage. Brundage, also on 1 September, had an ad in the *Solano Republican* offering the hunting public accommodations at the luxurious Teal, and a separate flyer made the point that "as the law permits shooting over land not cultivated nor inclosed...hunters need not pay any attention to parties claiming exclusive or other shooting rights or privileges." (*SFX*, 8 Sep 1893)

The cumulative effect of these moves on the Cordelia (and doubtless on the Ibis and Tule Belle as well) was catastrophic. One minute after midnight on 1 September, a horde of Suisun shooters burst onto the marsh and overran the clubs' leaseholds from end to end. (They encountered no resistance—the city hunters were not there and may not even have known of the early opener.) Through the month of September more or less continual gunfire over the marsh alarmed the ducks and degraded the traditional 1 October season opener for the urban club members. But for the horrified Cordelians, perhaps the worst news was the rumor that Mr. Brundage, who was not exactly a "manager" but the Teal's Afro-American steward, intended to bring "lewd members of both sexes to his arks." (Ibid)

No such improprieties occurred, but it took several years before the state supreme court invalidated the 1893 freedom-to-trespass statute. Meanwhile, the disgruntled Cordelians had to share their marsh with the Suisun general public and with other shooters from San Francisco. The encounters were not always polite.

The first confrontation occurred on 30 September 1893, when a party of 17 Suisun hunters was met at Teal Station by a steam launch with the Cordelia's feisty keeper, Captain Charles Chittenden, and two of the club's members aboard. Demonstratively sticking a Bowie knife in his belt, tucking a big pistol in his pocket, and seizing a loaded shotgun from outside his deckhouse, the good captain came ashore, announcing that he had heard some locals had arrived to "do him up" and he was "ready for blood." His employers, however, hastily rebottled him in the deckhouse, and soon all but the captain were in the clubhouse for one of Mr. Brundage's gourmet dinners. Once there, the Cordelia owners announced politely that they had rented the marsh as of 22 July 1893 and intended to keep all non-member hunters out. Equally politely, the Suisun delegation noted their rights under the new relaxed law regarding trespass, and, said the paper, the "matter was settled in a very satisfactory manner." For the Suisunites, it must have been: the paper reported that they all went shooting the next morning and did quite well. (*SR,* 6 Oct 1893)

Chittenden came ashore again after the meeting and apologized, but the locals were not inclined to be charitable. Only the week before, Chittenden had run some hunters off the property, seizing their guns and decoys. Referring to an unspecified recent "terrible tragedy," the *Republican's* editor predicted "very serious trouble" for Chittenden "if he points a gun at everybody he meets." (Ibid)

The paper failed to explain the "terrible tragedy," but it might have included a fatality from the kind of random keeper-poacher rifle fire that was seen in the mid-1880s. It was about at this time that the Tule Belle, which had built a fence to keep out poachers and stay within the definition of legally protectable land, lost its keeper to murder by a poacher, who was then acquitted in a local court on the grounds that he had acted in self defense. (*SFC,* 28 Oct 1893; McAllister, 1938) For whatever reason, local newspapers did not cover this killing. Little wonder that Chittenden had a chip on his shoulder, though one can question his good sense in confronting the entire Suisun party so belligerently.

Others sought to capitalize on the hostilities. We recall from Chapter II that it was during these times that Jabez Thickbroom lost his marshlands, probably from defaulting on his mortgage. The week after Brundage placed his ad in the *Solano Republican,* an ad appeared in the *Breeder and Sportsman* offering 600 acres for rent or sale of marsh "formerly used by the Tule Club, Baron von Schroeder, Mr. Jeslin (sic) and others." Those interested were advised to apply to Mr. "T. Hick Broom" at Teal Station. The following week the surname had changed to Brown, but thereafter the ad vanished, as did Mr. Thickbroom, who never again appeared in marsh annals. (Mr. Josselyn, who unquestionably was an avid reader of the *Breeder and Sportsman*, probably saw to that.) (*B & S,* 9 and 16 Sep 1893)

The Mallard Club

Matters became more complex when a group of wealthy but liberal San Francisco sportsmen under the leadership of Lloyd Eaton,

C. J. Dietz, W. Robertson, and Adolph Lorsbach formed a new hunting and fishing club on 13 December 1893. The new club declared it hoped to "visit the Teal Club on its inaugural shoot, which will take place on the 1st of January" (1894). The club piously commited itself to conservation with a rule that imposed a $13 fine on any member shooting more than 13 birds. (*B & S,* 16 Dec 1893)

At first, the new club, soon christened the Mallard, was popular in Suisun. Before the year was out, in November 1894, it leased some 1,572 acres north of the Teal/Cordelia marshes from George Tomasini at the incredibly low rent of $100 per year, or $.06 an acre. This was the same property that the Suisun Gun Club had intended to use. The sole lease signer was none of the known Mallard Club officers or even members, but a William Wattles of Alameda, apparently a figurehead.[5] There were other odd things about the Mallard, first and foremost that it was welcomed by the *Solano Republican,* even though it was made up of outsiders and was usurping the Suisun Gun Club's intended marshland.

The locals drew a distinction between the old-line Teal-Cordelia-Ibis-Tule Belle group on the one hand and the Mallard Club and freelance city hunters, who were not part of the "monied classes" (even though some Mallard leaders were), on the other. The former controlled their fiefdoms without contributing much to the community, whereas the latter, who did not have a staff of keepers and servants, patronized local guides, restaurants, and hotels. Suisun guide services offering not only good shooting but fine accommodations on floating clubhouses that became an important source of local income.

---

[5] William Wattles' name is only known through county records. He was never known to hunt the marsh. A Jack B. Wattles of Alameda was a Hardland Club founder in 1879, but he only lasted one year (possibly blackballed?) and never reappeared. The relation, if any, between the two Wattles is unknown.

The hosts didn't even need to forego their own hunting. As the *Republican* frankly observed, the guides could do most of the shooting and reap a double profit by selling the bag to the "city sports, many of whom are extremely indifferent marksmen." (*SR,* 31 Dec 1894)

The Teal Club principals, too, at first may have welcomed the Mallard Club as an ally against the Cordelia, but by the start of the 1894 season, they had soured on the idea. Having by then already settled their own quarrel with the Cordelia, they had no further use for rivals on their reacquired property. In fact, wrote the *Breeder and Sportsman,* "Now the Mallard and Teal are having at it 'hammer and tongs.' This is a disgrace to sportsmen." Like the Ibis, the Cordelia and Teal together by this time had thrown up a fence around some 3,000 of their shared acres. (*B & S,* 13 Oct 1894)

It did no good. Both the locals and Mallard Club members again infested the marshes. The Mallard moored its unnamed ark just 200 yards from the Teal's clubhouse and shot on the Teal's old territory for the entire 1894 season, despite being served with repeated injunctions to stop. (*SR*, 31 Dec 1894)

The hunting suffered from the overcrowding early in the 1894 season, especially when the northern migration was delayed by balmy weather. "The hunter is lucky who gets more than a mess," commented the *Breeder and Sportsman* disgustedly. (*B & S,* 17 Nov 1894) But then things improved, and in early December three Mallard members, including its president and vice president, killed 125 ducks (Whatever became of the club's 13-bird limit?)—and the president, Eaton, promptly found himself in court. Just the week before, the Cordelia's Charles Kellogg had filed legal complaints against a William King "and 39 others" and enjoined them against trespassing.

The Mallard Club party, for some reason without Dietz, was arraigned but quickly exonerated because they had arrived on the shooting grounds by slough. The King case (he was a local who presumably came by land) remained unresolved and dragged on into the next season, when it became the focal point of the legal battle. (*B & S,* 1 and 22 Dec 1894; *SFC*, 6 Dec 1895)

By 1895, the "unattached" San Francisco hunters were renting berths on arks hauled up from the lower bay such as the *Mary Joseph* and especially the *Crystal Palace*, which also anchored in Cordelia and Frank Horan sloughs. In an echo of the earlier maneuver of Suisun hunters to open the season early, the Mallard Club moved its ark to Espinosa Station in September and said it would start the season on 1 October—which was two weeks before the legal season began.[6] (*SR,* 4 Oct 1895)

Meanwhile, however, the Teal and Cordelia had taken matters into their own hands by building barricades across two tidal sloughs (the Sixth Reach and Frank Horan) that led to the shooting grounds. By this time the two clubs were essentially one, with Kellogg taking legal action in the name of the Teal, and some Teal members listed in Cordelia shooting parties. Before year's end, the two clubs had so closed ranks that they were called the "Cordelia Teal Club." (*B & S,* 19 Oct and 14 Dec 1895)

The reaction by both Mallard Club and local hunters to the slough barriers was unbridled fury, accompanied by anonymous threats to dynamite the obstructions. Both the Suisun and San Francisco press forecast new violence on the marsh. When it did not develop as predicted, press coverage was tinged with barely concealed regret. The *Breeder and Sportsman*, while admitting that

---

[6] Earlier in 1895 the Teal had lost a good deal of leverage over the Cordelia when the Espinosa (later called Drawbridge, then Cygnus) whistle-stop station was opened where the SP tracks crossed Cordelia Slough, thus avoiding the Teal Club blockade at Teal Station. By then, both clubs were fighting the common enemy, but it must have been comforting to the Cordelia to have a backup access.

THE TEAL CLUB BARRICADE AT THE MOUTH OF SIXTH-REACH SLOUGH.

The above cut appeared in the San Francisco Chronicle of Wednesday, and it is through the courtesy of the management of that great daily that the REPUBLICAN is permitted to reproduce it.

Range War episode. From the *Solano Republican* of 18 October 1895.

trouble had been averted on the Tuesday opener, found solace in predicting that there was a "good, fair chance of war on Sunday." (*B & S,* 19 Oct 1895)

There was another complicating factor. Although the Teal, Cordelia, and Ibis (and perhaps Tule Belle as well) by now had fenced and posted their properties, this no longer provided a legal bar to trespassing. In the February 1895 amendments to the game laws, salt water marsh land had been specifically exempted from such protection. This aroused the clubmen's anger, and they vowed to overturn the regulation at all costs.[7] (*B & S,* 16 Feb 1895, *SFC,* 16 Oct 1895)

As the 1895 season approached, the situation was quite volatile. On the eve of the opener, Mallard Club members knocked off the heavy padlock that secured the Teal barrier at the mouth of Frank Horan Slough, and, over the protests of the keepers, moved their ark to its previous year's mooring near the Teal clubhouse.[8] Came opening morning and Kellogg ordered off two trespassers, who simply ignored him. The only thing that probably prevented human bloodshed was the fact that the 1895 opening day was the best anyone had seen for years, with some 2,000 birds taken from the Teal, Cordelia, Ibis and Tule Belle properties. There were just too many ducks around to waste much time or ammunition on rivals, and, despite some crowding, everybody got enough shooting. (*B & S,* 19

---

[7] The club men probably considered it poetic justice that the first victim of the new law was Mr. Emeric, whose San Pablo Bay marshes were immediately and constantly thereafter invaded by eager trespassers until the law was finally set aside two years later.

[8] There was one report that the Teal then opened fire on the intruders and one keeper was seriously wounded or killed in the return fire, but this was never confirmed. It was either hushed up or, more likely, never happened. (*B & S,* 19 Oct 1895)

Oct 1895)

After opening day, the shooting went downhill as it always does, and the human conflict began to heat up in the courts, in the press, and on the marsh. The slough barriers remained in place, the trespassers kept on trespassing, and each side served injunctions on the other to cease and desist. Most observers felt the old clubs were wrong to erect the barriers, but opinion was about evenly split on whether they should be allowed to evict the trespassers. The press (especially the *Solano Republican*) did nothing to cool tempers with its rhetoric that "bombs of public indignation will explode" over the actions of the "despotic," "autocratic," and "aristocratic" clubs and their keepers ("lackeys"), who had accosted a local hunter on the Teal. There were thinly veiled threats of coming attacks against the keepers. (*SR,* 15 Nov 1895)

The local involved was a Mr. Hurley, who had been relieved of his decoys and his game before being evicted. He may have been singled out for harsh treatment because he was caught on a Monday, traditionally a non-shooting day, the violation of whose sanctity really upset the club men. To discourage off-day shooting, the keepers began following trespassers, spoiling their hunting by shooting blanks whenever birds came near, but this got expensive in ammuntion and employee time. They then set up removable, bright whirlygigs in the favorite ponds to scare off birds during non-shooting days. These proved effective until the trespassers began using them for target practice. (*SR,* 22 and 29 Nov 1895)

The Suisun hunters also retaliated against the gunfire ploy by similar shooting harassments of their own against the legitimate lessees. In a more aggressive step, they tried to provoke Captain Chittenden by coming out in force and shooting next to the *White Wings* on an off-day, daring him to oppose them. On this occasion, Chittenden used his head—he kept it down—but when the odds were not so unbalanced, he remained as obnoxiously courageous as ever. He infuriated local hunters by denying them a shortcut across a narrow arm of the Cordelia that might have saved them a five-mile row, and by threatening to chop up their beached boat if they ignored him. He also was accused of hiding barbed wire and, later, broken glass under tules along poachers' trails (though the first victims turned out to be his own underkeepers). (*SR,* 22 Nov and 27 Dec 1895; 4 Dec 1896)

Tensions increased later in 1895, after the Teal erected a 150-yard fence on either side of the Frank Horan Slough barrier and built a small hut for its two guards. Only a few weeks later, however, arsonists burned down the shack while the two guards were sleeping inside. The two barely escaped with their lives. This was in spite of the fact that the Teal had just taken on Suisun's constable, George Kinloch, as chief warden. In early January 1896, another fire—intentional or accidental—devastated the entire marshland from Cordelia to Teal Station. (*SR,* 22 Nov, and 27 Dec 1895, *B & S,* 4 Jan 1896)

By the end of December, 1895, the Mallard as a whole was starting to decline on the public relations front in San Francisco, though some individual members were still respected. A suit by the Ibis called them "an association of poachers... in an intoxicated condition, (who) have made life unbearable for anybody within their reach," and their arrogant behavior on the marsh, such as refusing to vacate club men's blinds when asked to, was drawing criticism. The club tried to better its image by listing some prominent figures as members (a state senator, a sheriff, a prominent attorney, and an assemblyman), but these were only "honorary," and were never seen on the marsh. The Teal's Mr. Brundage testified that the Mallard was "avowedly and confessedly a poaching organization," formed to shoot on others' land in defiance of the owners' wishes. (*SFX*, 29 Dec 1895, *B&S,* 28 Sep 1895, 4 Jan 1896)

By this time, Suisun hunters, too, should have been having serious second thoughts about the Mallard Club and the unattacheds. For one thing, it was reported that the Mallard had approached the

Cordelia with the proposition that they guard the club against trespassers in return for shooting privileges. Aside from being a thinly veiled protection racket ploy, it might have denied shooting rights to their allies in the Suisun freelance hunting community. (Kellogg turned them down flat.)

Furthermore, the Mallard was clearly the most obvious suspect in the guard-shack arson case (Suisun men were unlikely to have risked killing Kinloch), though no arrests were ever made, despite the Teal's offer of a $500 reward. And finally, the political clout of the Teal was demonstrated when it had Governor James Budd as a guest in late November 1895. But such was the depth of hatred against the old clubs that Suisun hunters volunteered to pay for one of two lawyers needed by the Mallard to defend itself against the Ibis suit, providing the Sportsmen's Protective Association (SPA) paid for the other. (A suggestion by the *Breeder and Sportsman* that the SPA also contribute to the Teal reward offer went unanswered.) (*SR,* 1 and 29 Nov 1895, *B & S,* 4 Jan 1896)

Legal Watersheds

The Mallard Club and its allies were heartened when Judge Buckles finally, on 6 December 1895, ruled in their favor in the case of William King, the trespasser arrested the year before on Cordelia Club land. The judge's questionable reasoning—among other dubious points that the plaintiffs' rental did not specify *exclusive* hunting rights and that damage to their shooting by the presence of uninvited hunters, especially at night, was merely an unproven allegation—was hailed as a judicial masterpiece by the *Republican.* Buckles might have done better to stand by the California legislature's law exempting saltwater marsh from trespass laws, but this act itself was being challenged as unconstitutional, and he chose a more devious route. (*SFC,* 16 Oct 1895; *SR,* 13 Dec 1895)

The happiness of the victors was short-lived. The ink was scarcely dry on the 1895 Buckles ruling before the Ibis Club circumvented it. Tired of running off Mallard Club and local poachers without any help from the local constabulary, the Ibis incorporated itself—in Colorado! This let the club call in federal marshalls for arrests that could be tried in federal courts.

At first the California State Attorney General loudly termed this a "federal fraud," but he fell quickly and totally silent when the Ibis Club's attorney, the appropriately named Charles Eells, cited his legal precedent: the incorporation in Kentucky some years before of the purely Californian Southern Pacific Railroad. (*SFC,* 29 Dec 1895) In those days, it was most unwise for anyone, of whatever rank, to challenge any legal precedent that benefited the SP. The Ibis suit against the Mallard noted on page 48 above (which it won) became the first legal test of that precedent.

And the Cordelia Club, though it had lost the King case in the Suisun court, immediately appealed it to the California Supreme Court, which not only overturned Buckles but in September threw out the law on the immunity of saltwater marshland from trespassing. (*SR,* 9 Oct 1896) It was just in time for duck season, but it did not quite yet put an end to the marsh war.

The War Winds Down

Just as in 1895, the 1896 season began with dire predictions of violence. There was even a specific threat by the Solano supervisors to close the duck season in the county for three years on 1 November. The local press was full of thunder against the King case decision. The clubs "will have trouble warding off the so-called trespassers. The fact of 49 or 50 rich men monopolizing the tule lands for their own sport and denying the privileges to the poorer classes is bound to cause trouble." "The establishment of preserves is English Lordism and is not in accord with American liberties and rights." (*SR,* 2 and 9 Oct 1896)

Just as before, both sides in the conflict ignored the recent legal decisions affecting the marsh. The Teal and Cordelia, under injunctions to remove their slough barriers, did nothing of the sort.[9] The Mallard Club moored its ark to one barrier, and on opening day its members (the ark held up to 28) crawled over it, dragging their boats behind them onto the forbidden territory. In December, when the "unattached" did the same, they were met by deputies handing out injunctions against trespassing. When ten of them continued, the deputies shrugged and told them not to stay more than 24 hours. (Given weather conditions on the marsh in December, the warning against overnighting was probably unnecessary.)(*B & S,* 7 Nov and 19 Dec 1896)

Not surprisingly, the violence was a self-fulfilling prophecy, especially after bluebird weather, a minimal duck flight, and overcrowding of the marsh resulted in a disappointing opener. On the famed Pringle ponds, "the hunters were so thick that it was decidedly dangerous to stand up, and a low-flying duck was comparatively safe, as no one dared shoot at it for fear of hitting his neighbor." There were at least 300 guns on the marsh, more than four times the reported number in 1895. (*B & S,* 24 Oct 1896)

By early November, there were reports of the Cordelia keepers firing a rifle at local trespassers, one shot holing a boat and another grazing the toe of a hunter's boot. The locals departed, vowing to return and exact revenge if they were fired on again. In December, there were two more cases of arson, one again involving a guard shack at a barrier and the other a keeper's cabin, where ammunition exploded dramatically as the house burned to the ground. A local trespasser expelled from the Cordelia by Captain Chittenden

---

[9] A year later they were still there, and it is unknown when they were eventually removed. Freedom to hunt and fish navigable sloughs, regardless of who owned riparian rights, was only firmly and forever established by the California Supreme Court in 1912. (*SR*, 9 Oct 1912)

had the evicter arrested for assault with a deadly weapon and "menacing language," but secured no conviction. Perhaps more troubling to the locals was the fact that irresponsible trespassers, whether locals or "unattacheds," had been shooting at cattle, forcing their owner to post his Joice Island property. (*B & S,* 7 Nov and 19 Dec 1896; *SR,* 25 Dec 1896)

But as for the threat of shutting down the season ... well, this was not unlike my father forgetfully volunteering to teach Sunday school during the duck season. When principles confront reality, the costs can get get prohibitive.

By the end of the year, the fight had gone out of the "unattacheds" and the Mallards. The Sportsmen's Protective Association "began the new year by giving up the fight against the private clubs." First the *Crystal Palace* and then the Mallard scow quietly slipped their moorings and vanished down Cordelia Slough, never to return. Only two years into a five-year lease, the Mallard simply abandoned its Tomasini tract, leaving only three of its leading members— Dietz, Eaton, and Robert Boyer—to become outsider members of the Suisun Gun Club that succeeded it. The Sportsmen's Protective Association, which had strongly supported the propaganda campaign against the private clubs, advised its members to "dig down in their pockets and lease, as others have done, a tract of land for hunting." (CR; *B & S,* 19 Dec 1896; 9 Jan 1897)

Peace at Last

Like all wars, this one left some permanent traces. One probable legacy of the Teal/Cordelia conflict that helped start the fireworks endures to this day. The Cordelia, Ibis, and some other owners still pay taxes on tiny pieces of property along the Southern Pacific tracks near the now defunct Cygnus and Teal stations. The most likely explanation for these otherwise unexplained mini-plots is that they were bought as insurance against the Teal or any other club ever again sealing off rail access to the marsh. (CR)

In the shorter run, the *Solano Republican* imposed a virtual news blackout on the "gentlemen's clubs" (Teal, Cordelia, Ibis, Tule Belle, and later the Joice Island) from the late 1890s until 1905. This probably suited the intended victims just fine. They seem to have wanted only to be left alone, and they may even have enlisted some railroad friends to help. A 64-page Southern Pacific promotional booklet published in 1896, entitled *California Game "Marked Down"*, describes dozens of hunting and fishing sites along rail routes throughout California—but gives not a single word about its marsh stations near Suisun or about the surrounding wetlands. Nor is Suisun mentioned in other literature about a special oversized caboose that could be rented to carry parties of up to 50 hunters, their gear, and their dogs (fare for a dog: $.20) anywhere the railroad ran. If anything, the pamphlet may have helped divert hunters' ambitions *away* from the Suisun Marsh—which may have been part of its purpose.

By early 1897, the local Suisun shooters had banded together to lease their own hunting lands next to the existing preserves and thus prevent further encroachment by the outsiders. The Mallard marshes were bequeathed to the Suisun Gun Club (SGC), which added to the leasehold and opened the next season with 1,897 acres. Ironically, in view of the *Solano Republican's* former fulminations against "aristocrats," the SGC leaders were themselves a kind of local aristocracy, with Suisun Council member William Bryan as president and Under Sheriff Thomas L. Robinson as secretary, and most other known members holding influential positions in Suisun and/or Solano County society. Suddenly, the local paper's prejudice in favor of trespassing withered. Opening day on SGC property, noted the *Republican* approvingly, would be for members only, and thereafter each member would be allowed only one guest a day, someone who would have to "come from a distance." (*SR,* 8 Jan and 24 Sep 1897)

But the most startling sea change came in November, when

the *Republican* editorial, in a rare mention of its old adversary, actually *praised* Captain Chittenden's "polite" expulsion of trespassers from the Teal Club, the first nice thing the paper had ever said about him. Having failed in their fight with the "sporting aristocrats," the publishers seemed to be on the point of joining them. (*SR,* 26 Nov 1897)

End-of-the-Century Situation

In 1897, the Cordelia again renewed the lease for all the Chamberlain Estate marsh except that reserved for the Ibis Club and the railroad. Although carried only in the Cordelia name, the lease was also the Teal Club's, which continued to shoot its traditional ponds.

The Chamberlain heirs were growing up. Emily had married and moved east, Freddy had gone off to Yale and then to some giddy (presumably) postgraduate activities in Paris, and Mary had taken a social leap into nobility as the Baroness von Seelhorst of Baden Baden, Germany.[10] Their Suisun estate, termed Swamp and Overflow Lands by government surveyors, probably did not rank high in their interests if, indeed, it ever had. (CR, affidavits for leases)

The years 1898 and 1899 saw a swift sprouting of new clubs, organized by both locals and outsiders. After only one year, the Suisun Gun Club split up over a disagreement about funding. The original club wanted to charge outsiders $5 a day, whereas another group wanted to make guest privileges free. The latter broke away and formed the Pringle Gun Club that took in Mr. Peyton's famous Pringle ponds. Three days a week were reserved for the eight members, three days were closed to all hunting, and the seventh day

---

[10] Coincidentally another German baron, J. H. von Schroeder, who *also* was married to a Mary, was an owner of the Tule Belle in 1901. The connection, if any, between the two noblemen is unknown.

would be open to visiting hunters. (Note the parallel with the Wisconsin experiment in permit hunting.) Among its members were some San Franciscans, including the first corporate member, Golcher and Co.[11] Not only the Pringle and SGC but the newly formed Potrero Club, located on the "Lang and Morrill marshes" near Denverton, with headquarters on the Hastings Ranch, was taking in San Francisco members. (*B & S,* 8 Jan 1898; *SR,* 6 Oct 1899)

Another new hunting group was the Olympic Gun Club (an offshoot of San Francisco's Olympic Club and a year later to be renamed the Field and Tule Club), which established itself at the northwest corner of the marsh, taking in today's North End and Whistler clubs and probably other wetlands that have long since vanished. A Canvasback Club (by 1905 renamed the Family Club) was formed near Goodyear Station, and a San Francisco Club, consisting entirely of San Francisco *Chronicle* employees and executives, was bargaining hard for the famous Carpenter ponds near Denverton as the century drew to a close. (*SR,* 27 Oct, 1and 8 Dec 1899)

Old-style trespassing was not yet entirely dead, however. Fittingly, the last publicized case in the 19th century involved that old Mallard hand Robert Boyer, who in 1898 was let off by Judge Buckles with a small fine and a warning not to sin again after being snagged on Cordelia property by three keepers. Boyer's defense—that he simply did not know quite where he was on the marsh at the time he was caught—was questionable, considering how long he had been shooting there. (*SR,* 16 Dec 1898)

As for the *Republican*, one of its publishers showed that he wasn't prepared exactly to *join* the aristocrats. He was just preparing to siphon off some of their wealth.

---

[11] Alas, the heir to this famous old gunsmithing establishment, from whom I made my first shotgun purchase in 1942, went out of business and joined Abercrombie & Fitch around 1950.

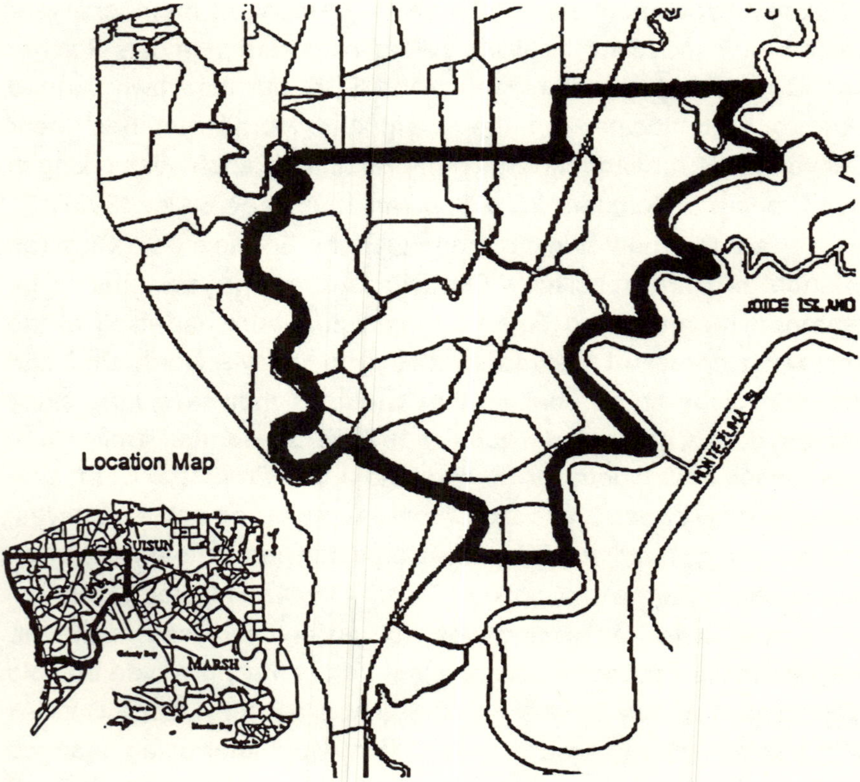

CHAMBERLAIN ESTATE

The estate, obtained by a patent in 1871, comprised more than 5,000 acres of prime marsh land. Shown here are the duck club properties that developed in and around the estate after it was sold in 1905.

CHAPTER 4

WEST-SIDE GOLDEN AGE (1900-1935)

"In their bowler hats and sidewhiskers, and laden with jugs of whiskey and 10-gauge Parker shotguns, the club hunters crossed on the ferry from San Francisco to Oakland and boarded the steam cars, to debark at Teal, and Cygnus, and Jacksnipe station for a leisurely weekend of duck hunting."[1] (SSCD.1, p. C-9)

For a while after the turn of the century it appeared that resident-nonresident hostilities like those of the mid-1890s might resume on the marsh. That did not happen, perhaps in part because duck hunters who came from the western counties gradually switched

---

[1] My severest critic, Dick Dinkelspiel (see p. 61), has taken me to task for this quote. He says that in the early decades of this century bowler hats and sidewhiskers were long out of style, that 10-gauge guns were unnecessary for yesterday's easy-decoying ducks (the only time he saw 10 gauges "were those used by the market gunners... whom I usually sent to jail for 6 months—the maximum"), and that boozing wasn't that big a deal ("most gunners were sober—well, at least by shooting time"). The whole package, he says, adds up to a picture of rich wastrels, "yet these were the people who saved the marsh for us." That is certainly so—but, absent maybe the Parkers and bowler hats, don't all of us today know rich wastrels, some of them whiskered, who are devoted to saving the marsh?

over from leasing to owning their property. They were still absentees for most of the year, but somehow ownership seems to have bestowed more respectability on them in the locals' eyes, and it may have given them a new sense of stewardship for the land and its game. Either way, the 20th century ushered in what has been termed the Golden Age of hunting on the marsh.

Infiltration from the West

The 1900-1901 season was a banner year. The average bags may not have approached the record individual scores that some gunners had turned in during the 1880s and early 1890s, but opening day saw 25 guns on the best west-side ponds pull down more than 1,250 birds, for an average of more than 50 each. (*B & S,* 6 Oct 1900) With that kind of productivity, no one could complain.

Nevertheless, at the end of the 1900-1901 season, the famous Pringle Ponds were pulled away from the locally-run Pringle Gun Club and leased to a seven-man San Francisco syndicate. "These ponds have been under the control of Suisun sportsmen...for the last forty years and this deal closes all available hunting grounds to local hunters," grumbled the *Republican,* with understandable resentment but less than total honesty. (*SR*, 4 Jan 1901) For the previous two years the Pringle Pond lessees had been a mixed Suisun-San Francisco group that had not welcomed trespassers from whatever quarter. Still available to Suisunites were considerable areas of unclaimed (or at least unrestricted) marshland, the maze of sloughs, and, of course, membership in the Suisun Gun Club.

Still, the continued encroachment could not have been welcome in the community, especially when the 1901 season opened disappointingly. This was the first season where a daily maximum (50) was set on the bag, but "No Effort Was Required to Keep Inside the Limit," was the dry *Republican* leader on October 4. In the story that followed, the paper noted the modest bags that locals had made (roughly 20 to 40 for the good hunters) and remarked smugly that the five Pringle shooters from San Francisco had overslept and only bagged 38 birds among them. (*SR,* 4 Oct 1901)

But the vindictive, contemptuous tone of earlier reporting on the city slicker hunters was missing. The paper gave the names and occupations of the seven new Pringle owners in some approving detail, including that of one Frank Maskey, and from then on it covered the Pringle shoots in a positive, sympathetic manner.

Yet another piece of top-quality marshland, the Stewart ponds off to the east, was leased to outside invaders in November 1902 "for a number of years," and again the *Solano Republican* coverage was coolly objective. This time only two men, both San Franciscans, were involved in the takeover. One of them was called J. Downey Harvey. (*SR,* 21 Nov 1902)

From then on, new clubs sprang up on the marsh like pasture mushrooms. The Suisun Gun Club lost members, dropping down from 25 at its birth to six in 1905, but its founding fathers and other eager hunters were busy putting new ones in business: the Beetville Shooting Club near Cordelia in 1901, the Tule Gun Club in 1902 (later merged with the Tule Belle), the Armijo, Montezuma, King, and Oakland in 1903, the Belvedere (later renamed the Volante) and Cotati in 1904, and the Harvey, Big Four, Family (formerly Canvasback), Red Head, and Badgers in 1905. By 1908 there was a Marsh Club and Morrow Gun Club (probably the forerunner of today's Morrow Island Club), and, in 1909, a Clover Leaf Club near Cordelia. There was even a self-styled Piker Club, made up of those who had no property rights but shot from railway embankments and public roads. Of these, only the Family and Montezuma still survive under their original names, and the Montezuma has moved at least twice from its original site near Collinsville. (*B & S,* 17 Aug 1901, 17 Oct 1903, 31 Oct 1908, 25 Dec 1909)

Meanwhile the *Republican*'s former favorite whipping boy, the Cordelia Club, completely vanished from the paper's columns, as did the Teal, Ibis, and next-door Tule Belle, all clubs with fine shooting and colorful characters. After every opening day from 1897 through 1904, the *Republican* would report in careful detail all the other areas shot, the scores of individual hunters (both local and out-of-town), humorous asides, and the old or new clubs involved. As the season progressed, new reports followed almost weekly in most years. But the Cordelia and Teal got only one passing mention each during this span, and the Ibis and Tule Belle were ignored completely. The news blackout even persisted when Freddie Chamberlain apparently took the unusual step in 1902 of leasing the entire family estate for a term of five years to the Ibis Club, thus potentially freezing out both the Teal *and* the Cordelia. (McAllister scrapbook)[2]

Perhaps this reporting vacuum was merely to inflict journalism's worst punishment, anonymity, on the intruders, but more commercial considerations probably dominated. The real estate deal of the century—as far as the marsh was concerned—was brewing in the background, and those involved had a vested interest in keeping outside interest in the areas under negotiation to a minimum.

The Syndicate Strikes

The buyers that joined forces to strike the deal were an unlikely syndicate even for turn-of-the-century California: a pillar of Suisun society, a San Francisco candy-maker, an Irish immigrant, and, in the background, a professional gambler and a dead man. In those days, as today, people involved in duck hunting seemed to maintain peculiar associations.

On August 4, 1905, Edward Dinkelspiel, Frank Maskey, and Joseph Harvey were officially recorded as the new legal owners of the entire 5,332-acre Chamberlain estate. They took possession for $110,000 in their own names and the names of their wives.

---

[2] Another victim of the blackout was one of the finest "new" clubs, Joice Island, founded in 1902, when Arthur Goodall bought its marsh from the Rush family. (CR)

(Dinkelspiel papers) Later, it emerged that two others, Frank Daroux and Charles Fair had also contributed financing. The sale made provision for only two leases: that of Harry Babcock and M. Hall McAllister, who along with the now adult Freddy Chamberlain retained the Ibis lease and controlled the Teal and Cordelia subleases, and an unspecified part of the estate, probably the drier northern upland, that Freddy shared with William Pierce. (CR)

The record may say 1905, but other evidence indicates that the deal probably began to solidify around 1902 and perhaps earlier. Whatever the precise starting date for negotiations, the sale, once consummated, foresaw all lessees being permitted to remain on the marsh until their leases ran out, "when the new owners will decide what to do." That intention, to the undoubted consternation of the duck club owners, was reported at the time as being the draining and cultivation of the marsh for asparagus.(*B&S*, 17 Jun 1905)

But even before the leases ran out, the new owners began quietly selling choice bits of marsh not to farmers but to hunters—including their own syndicate members Maskey and Harvey. Whether or not they ever had any intention of selling to farmers or whether this story was designed merely to hold up land prices is unknown, but the same story about probable development was floated when they were selling the last of the estate in 1911 (see p. 93).

Let's take a closer look at those who engineered the deal.

Edward Dinkelspiel

Edward Dinkelspiel, the leader, we already know from his newspaper's coverage of hunting in the area and his now-muted crusade against the rich absentee lessees. But he was far more than just an editor of a relatively obscure paper. He was born the son of an 1857 immigrant from Bavaria, Moses Dinkelspiel, who had worked his way up from hardware clerk to head of the Solano Board of Supervisors, bank director, and a director of the state university. For his part, Edward was not only an editor but a local lawyer, judge, and insurance man in Suisun, where he spent his entire life (1862-1937). He was an avid duck hunter in his youth, and County records show that in his adult years he was intensely involved in real estate. The family was such a bulwark of respectability in Suisun that when Edward died in 1937, the county authorities passed his judgeship to his young son Richard as a matter of course—just as the heir graduated from law school and before he had even passed his bar exams.

"Colonel" Frank Maskey

By shopping for the best ingredients at 4:00 every morning, Frank Maskey had singlehandedly turned his humble San Francisco candy store into a city institution. It became such a landmark that when construction of a new bank threatened its destruction in the 1980s, the city refused to permit the old store's demolition. (The building, now a copy shop, is at 66 Kearney Street, where Frank's old trademark, a mask and keys, still peers out through the grime of neglected stained-glass windows on the second floor.) At the turn of the century, no proper San Franciscan would let a Christmas or Valentine's day go by without bestowing at least one five-pound box of Maskey's chocolates on some business or social favorite.

Although we don't know who specifically introduced Frank to the marsh, it was probably Joe Harvey. As for his assumed rank of colonel, that was perhaps the result of a *Republican* typesetter's error in ascribing the same rank to Maskey as that held by Colonel Preston, a fellow lessee of the Pringle Club in 1901. Once that mistake had been made, in a 1902 column, Frank was never allowed to shake the title whenever his name appeared in the *Republican*. Given Dinkelspiel's wry sense of humor, it is mildly surprising that Maskey was not promoted to general when World War I broke out.[3]

---

[3] *The Breeder and Sportsman* referred to a "Colonel" Dinkelspiel in an 8 Oct 1910 article on duck shooting. Maskey's revenge?

Joseph D. Harvey

The *San Francisco Call*'s headline reporting Harvey's death in 1907 from pneumonia referred to him as a "Prominent Turfman and Public Character," hailing his generosity in giving away the riches he had made on the race track and in real estate. Having arrived in California in the 1860s as a penniless lad from County Tyrone, Ireland, he must indeed have had a golden touch, for by the time he died at age 52 he owned three champion racing mares and had an estate estimated at $1,000,000. (*SF Call,* 10 Aug 1907) Eighteen months earlier and far less flatteringly, the *Call* had identified him as the co-proprietor of the Green Palace, a San Francisco pool hall and gambling den that had been raided despite the owners' "substantial prior payoffs to police at all levels." The paper reported that Joe had smiled benignly as police bundled more than 250 customers and employees off to the hoosegow, but that that smile "boded no good for somebody." (*SF Call,* 28 Feb 1906)

Again, it is not certain when Harvey first visited the Suisun Marsh, but he is known to have been a close hunting partner of Charles L. Fair by 1901, when the pair of them used Fair's steam yacht *Lucerno* as a moored clubhouse for themselves and their guests in some unnamed slough south of Cordelia. (*B & S,* 4 Oct 1902) By October 1902, he was a member of the Tule Belle, where he was named in a lawsuit against the club and he was also one of the new lessees of the Stewart ponds.(CR; *SR*, 21 Nov 1902) The marsh shot by Joe and Charles was probably the same property that Joe later owned next to the Tule Belle and near the confluence of the Suisun and Cordelia sloughs, where the Miramonte and Antioch Goldeneye clubs are now found. By October 1905, when the *Republican* suddenly renewed its descriptions of properties in that area, the clubhouses on the Harvey and Tule Belle clubs (the latter then owned by Hermann Oelrichs) were referred to as the best equipped and most lavishly furnished on the whole marsh.(*SR*, 20 Oct 1905)

Frank Daroux

Joe Harvey was the patron and probably the protector of a young professional gambler, Frank Daroux, who was his partner in the raided pool hall. After Joe's death, Frank came into the news more and more often as police gave him one warning after another or simply raided his establishments. His betting covered everything from horses to would-be governors, and he stood bail for at least one bunco artist, but he was also invited to a reception by Henry Ford and was accepted indulgently by San Francisco's laissez-faire society.

Frank came rarely if ever to the marsh and is best known for his activities elsewhere. In 1909 he married Tessie Wall née Donohue, who ran what the papers called a "disorderly house" on O'Farrell Street—the politer papers called it a "resort"—and was the acknowledged queen of the city's madams. Her house was decorated with ornate cast-offs from the wealthiest homes for miles around in a hideous mismatch of styles that led one sophisticated Gumps salesman to comment that "It isn't that Tessie lacked taste—she had too much of it." Displayed prominently in her boudoir was the embroidered observation "If every man was as true to his country as he is to his wife, God help the USA." (Moffat, pp. 175, 176)

As confirmed social rebels, Frank and Tessie felt mildly ashamed of having legalized their relationship, but Tessie made up for it in her own way. In 1917, right after they split up and Frank took up with another lady, Tessie shot him down on Eddy Street, an incident paralleled in that old hit song "Frankie and Johnny," even unto the "Root-a-toot-toot, three times did she shoot" part. Unlike the ballad's victim, Daroux survived his wounds. Tessie wasn't all that bad a shot—she got him twice in the lungs and once in the liver—but Frank was obviously a tough egg. Afterwards, however, San Francisco must not have seemed the same to him, for he left the city and died in New York in 1928.[4] (*SFX,* 19-21 Dec 1917; 12 Mar 1928)

Charles L. Fair

And then there was that most silent of partners, the dead man. At the turn of the century, when he was only about 33, Charlie Fair was already very rich, having bought up about $3,000,000 worth of California real estate, much of it wetlands in Sonoma, Napa, Yolo, and Solano counties.

Another close friend of Joe Harvey, Fair must have entrusted his share of the money for the Chamberlain deal to the other syndicate members very early. On August 14, 1902, three years before the property officially changed hands, he and his wife were killed together in an automobile crash in France, surely among the first trend-setters in what would later become a standard French and Californian way of checking out.

Interestingly, the settlement of Fair's estate in 1910 showed that he was also connected with Frank Daroux, having given him a promissory note for $2,000 on March 3, 1902 and depositing 1,300 shares of Sacramento Oil stock as collateral for it. (Small potatoes for Fair—he was probably playing poker with Daroux and tried to win with a high two pair against three of a kind.)

All in all, it was a unique combination of the respectable, the disreputable, and the other-worldly, something that only the easygoing California social climate of the times could have spawned. Whatever they may have thought of each other in private, the partners put up $35,000 in cash in 1905 and in 1906 owed the estate only $47,860 of the original $75,000 balance. (1906 Solano tax book)

The Great Sell-Off

Once the Chamberlain estate had changed hands, there was

---

[4] In the song, of course, Frankie was the lady and Johnnie her john, but that's San Francisco for you.

no longer any reason for the *Solano Republican* to keep silent about the rich men's clubs. Quite the contrary, the paper's editor now had a vested interest in touting their impressive shooting records and other attractions as a way of driving up the price of real estate; he and the other new owners wanted to turn a profit on their investment.

Nevertheless, some hostility, especially regarding the Cordelia Club, still remained. When the Ibis/Teal/Cordelia lease ran out in 1907, the first two clubs remained tenants of the new owners while dickering to buy the leased properties. The Cordelia, however, merely shifted operations outside the old estate to the west of Cordelia Slough, where it expelled a group of Vallejo hunters by offering the absentee owner, an Englishman named Cook, a higher lease. (Shades of Cordelia's great 1893 Teal robbery.) The old Cook Club lease, however, was for a full year and had commenced on October 15, 1906, the beginning of the 1906 season. But in 1907 the season began on October 1, and, despite outraged protests by Cordelia's President Kellogg, the old lessees kept their shooting rights for the first two weeks of the new season. The *Solano Republican's* coverage of these developments showed little sympathy for the plaintiffs, and one can almost hear the Teal Club's laughter. (*SR,* 4 Oct 1907)

As soon as the new owners had bought out the Chamberlains, they began swapping pieces of the territory among themselves, as well as selling to outsiders. Even some of their wives began staking out claims to cerain tracts. Their immediate goal seems to have been to recoup their individual investments by selling off some choice pieces of hunting territory first, and then waiting for the rest to appreciate in value. That, at least, is how the selling pattern developed.

In 1902, the Teal was the first club on the marsh to become a "company" (although it only became legally incorporated in 1908), and as a company became the first, also in 1902, to pay personal property taxes. The Teal is next recorded as having passed from Maskey to Louis Titus and Wickham Havens in September 1907, but this transfer was clearly in the mill at least a year earlier. A May 1906 letter to Maskey from Titus promised a $2,000 payment in June but, noting that the complete annual payment would be "somewhat difficult" just then, asked for a delay until December "on account of recent events"—undoubtedly the 1906 San Francisco earthquake and fire, which had burned out his office just a month before. (Stiff-upper-lip understatement is not a British monopoly.) (Dinkelspiel papers)

Dinkelspiel and his partners made a 100 percent profit on the Teal deal, and other sales recorded in 1906-1907 probably did equally well for them. Some 1,300 acres of prime duck hunting area, encompassing the Allegre Club (Achille Roos), Green Lodge (Bill Richards again), Jacksnipe (William C. Murdock, James K. Prior, Jr., and James Rolph), and Ibis (Judge Henshaw, Henry Pfortman, and Ed Graney) were officially deeded over to their new owners. As for the Cordelia marshes, they were first leased to local gunners and then sold to the Teal's Titus and Havens, who promptly leased them back to the Cordelia's old members. Assuming the rent the Teal asked was not exorbitant, this probably helped settle the old feud between the two clubs.

In 1908, John F. Seymour and nine others first leased and then bought the Seymour Club, a stretch of marsh north of the Cordelia Club that had formerly been owned jointly by Harvey and Fair. Only a year later, just two of these were still there, a William C. Murray having bought out all the rest, including Seymour himself. Other sales by the Dinkelspiel-Maskey group in 1910 were to the Presley Company (for what became the Sheldrake Club) and to E. E. Brownell for the southernmost parts of the tract, which was Joe Harvey's property (the "Hawley" Club in the deed book). (CR)

Wickham Havens, a San Francisco blue-blood and real estate entrepreneur, must have been kicking himself for not having bought the Chamberlain estate himself. To have been outmaneuvered by the syndicate, of whom only the hidden partner Fair was socially exalted enough to be listed in the San Francisco *Blue Book*, and to

Members of the Volante Club pose for the cover of the November, 1911, issue of *Western Field*. The water spaniel shown here was a standard retriever in those days.

be forced to pay twice the price for the Teal that the upstarts had just paid for it must have been especially galling. We need not waste too much pity on him, however. His ad in the 1900 *Blue Book* reveals him to be an original (and probably the major) developer of Piedmont, one of the fattest of the Bay Area's fat-cat suburbs.

As for Titus, he may have been late to the starting gate, but he made up for it by snapping up whatever Suisun marshland came on the market. Although he is only known to have shot at the Teal, where he shared ownership with Havens, Ira Miller, Edwin Goodall, and Walter H. Leimert, by 1910 he had bought the entire Cordelia Club from the Dinkelspiel group, and, with two other partners, the northern half of Joice Island except for the Volante Club. He soon sold the Cordelia to James Otis and part of the Joice Island tract to Patrick Calhoun, whom we met in Chapter I and will soon encounter again. (CR)

Although the records are strangely silent on the sale prices, Titus probably did not lose on them. John De Witt, writing in the *Overland Monthly* in 1910, reported that "duck fever" had swept over the business community in the past few years, overtaking tennis, billiards, golf, and bowling as a recreational fad. The boom had doubled the number of duck clubs and led hundreds of businessmen who had not touched a gun for years to renew their interest in hunting. (De Witt, p. 439)

De Witt's article itself must have spurred interest. His listing of prominent San Franciscans who hunted the various clubs could only have added to the feeding frenzy of those wanting to get a piece of the action. One of these was a young and none-too-wealthy San Francisco lawyer who in 1912 joined his law partner William Denman in the purchase of 443.7 acres across the tracks from the Teal Club for $33,375.

The young man had no business throwing that kind of money around. I know. He was my father.

High-Life Shooting

As noted in the previous chapter, the early duck clubs were run by the cutting edge of society. By the turn of the century many of the original founders had died off, including all of the known original Teal Club team, but some of their practices remained as legacies.

Following in the partying tradition Commander Floyd started at the Teal Club, Herman Oelrichs, "the well-known sportsman from New York" who arrived in California in 1894, bought the Tule Belle in 1903 and immediately began operating in style. In 1903, before lavish clubhouses were the mode, Oelrichs valued his "improvements" at $3,000. "One could go out to the blinds in a dress suit and pumps, shoot ducks, and come back to the club-house without a change of raiment being necessary, so convenient and handy were the appointments." McAllister confirms from his own personal participation in "enjoyable outings" with Oelrichs: "We always went out and returned on special trains and there was no stint of the good things of life." (McAllister,1930 p. 285)[5]

After Oelrichs' death aboard a transatlantic liner in 1906, Edward H. Harriman bought the club from Oelrich's estate for his son, but he owned it only a few years before apparently abandoning the property. It went downhill fast, degenerating into a market hunter's domain, where night shooting was so incessant that it affected all the surrounding clubs. Eventually, about 1910, it was put up for auction and bought by two enthusiasts, Guy Earl and Amby Buckley, who shot it for several seasons. (*B & S,* 29 Nov 1913)

As we have seen, not all the clubs were owned by San Francisco and East Bay elites. The Suisun Gun Club, which was founded in early 1897, had 17 members in 1902, only four of whom

---

[5] Hmmm. One senses a certain Victorian reticence here—were there any other good things left undescribed, Mr. McAllister? Mr. Oelrichs had the reputation of being a no-holds-barred entertainer.

came from San Francisco. All but one of the others were Suisunites.

Nor were all the new owners on the marsh pillars of propriety, though a good many were very rich and influential. Patrick Calhoun, who in 1908 arranged to buy most of the northern part of Joice Island from Messrs Titus, Spring, and Hotchkiss, on 25 May 1907 was indicted by a Grand Jury for bribery in his mismanagement of San Francisco's public transport system. The accusations of graft and other corruption were impressive enough that the New York National Bank on 2 June that year refused to buy any more San Francisco bonds. (The class-conscious De Witt, not quite daring to ignore Calhoun yet not wishing to lend him prestige, deftly skirts the issue with one very brief mention of his club sandwiched between detailed descriptions of the neighboring Joice Island and Volante clubs.) Later, from 1913 to 1915, Calhoun was often in the papers for various financial misdeeds, including the real estate scam at the Collinsville Cut (Chapter I). And he never did pay for his duck club, which John Spring had to get a court order to repossess in 1916.

Ah, those were the years in San Francisco. Its high society was just as hard to break into as any on the east coast, but it is doubtful that Boston's Cabot or Lowell matrons were ever heard to sigh, as did Alma Spreckels to a companion at the St. Francis Hotel on receiving a cool nod from a pair of younger women, "Those de Young girls are nice, but we've never been close since my husband shot their father." (Moffat, p. 179)

The Slumping 'Teens

The Dinkelspiel syndicate timed its sales well, for the 'teens of this century saw a definite slump in duck shooting, and enthusiasm for hunting fell. Though there were a few red-letter days in November 1911, heavy flooding spoiled the later shooting. From the end of December, the late Joe Harvey's club was "out of commission" for the rest of the season, an even clubs reportedly "bursting" with ducks on their ponds had no hunters at all on the last Wednesday of the year.

(*B & S,* 30 Dec 1911)

As the years passed, the new generation of owners also grew old and died or sold out. Cordelia's crusty old Kellogg expired in 1913, the same year that local Suisunites bought out the San Francisco members of the Pringle Ponds Club. Bill Bradford died in 1916, and Frank Maskey in 1917. Also in 1917, Leimert and Havens split up their partnership in San Francisco, and the Teal Club ownership passed to Edward I. de Laveaga, a prosperous landowner in Orinda. By the 1919 season, de Laveaga had posted his own map and regulations in the Teal clubhouse, where they still remain.

Next door and across the tracks, G. S. Arnold bought out his law partner Denman in 1919, with a final payment made in 1922. Although his regular hunting companions helped pay operational expenses, the property remained in his name from then until his death in early 1942. Today, the young men who bought the place then are still the owners, if not quite so young.

Down the tracks to the south, the tightly integrated Roos family was another single-owner entity which would continue in the same family over a span of five decades. From 1908 onward, Achille Roos, like Oelrichs before him, valued his clubhouse at $3,000 in the tax books and cheerfully showed off his wealth. (A favorite Arnold Club story from the 20s reports a plaintive, accented cry heard from next door, "Achille, Achille, take your hand off da blind—da diamont glittahs too much!")

But the Golden Age was in decline. The first years of legal limits, starting in 1901, had seen many hunters pick up their 50 birds on good days, but by the time the limit dropped to 35 in 1907, no one on the Suisun Marsh limited on the opener. Two years later it was 25, and for many hunters, it could have been ten or 15 without cramping their style. A severe drought in 1912 led to average bags on opening day of only three or four birds, and 1913 was another poor year. Dry years returned in 1917, accompanied by fresh-water diversion up the Sacramento and San Joaquin rivers, and leading to salt water intrusions ever further up the delta. Uncontrolled effluents in the river systems and San Francisco Bay polluted habitat. Interrupted by brief resurgences, the duck populations went down year by year. ("Docks, docks," grumbled Achille Roos to my father during one bad year, "So don't talk to me about docks!") Talk or not, though, Achille kept his club until he died, as did my father and most of the other owners on the marsh.

The shooting may have declined, but in the San Francisco business world the clout of some club owners remained formidable. In 1929 a work map of the Southern Pacific showed where some 550 feet of trestle near Teal Station had been replaced by fill "at the request of the Teal Gun Club," perhaps to let the members sleep better. This, however, had apparently caused some siltation in Frank Horan Slough, and the present work, again at the request of the club, was to dig a new channel under the right-of-way to permit better water circulation through the slough.

The closing of this era was notable for the establishment of California's first state-owned game refuge, Joice Island, where 1,728.4 acres had passed from Pat Calhoun to John Spring to Andrew Mahoney, a cattle rancher, in 1916, and from there to the state in 1931.

Bill Richards, known in the 1930s as the "Duke of Cordelia" or "the Baron of Suisun", and his Green Lodge (nicknamed "The Limit") were still the toast of the hunting community. The Cordelia Club, now owned by James Otis and later by his estate, soldiered on. The Teal Club was taken over in the 1930s by Burlington M. Carlisle, owner of a San Francisco printing firm and a Teal member since 1924, who (doubtless with the approval of Commander Floyd's ghost) reestablished the club's reputation for lavish entertaining.

During the 30s, most of the last surviving members of the early clubs died off, most notably Captain Chittenden, Bill Richards, and Charlie Eells. Only the faithful chronicler M. Hall McAllister carried on, dying in 1948 at the age of 87.

They were boisterous years, those first decades of this century, marked in part by our continued heedless exploitation of the environment but also by the beginning of conservation-mindedness. Though often depicted as mindless mass slaughterers of ducks, the clubs and their members did more to preserve the waterfowl resource than they are given credit for. Through their baiting, their confining of shooting to two days a week, and their overall observance of hunting regulations, they probably justified the judgment passed on them by *The Breeder and Sportsman* as early as 1909: "Duck clubs (are) the greatest protection agent that has ever been exerted in behalf of the birds." (*B & S,* 2 Jan 1909)

Bill Richards' Green Lodge ("Ain't it 'The Limit!'") and his duck-hunting wife and guest in 1906 (*Western Field,* No. 6, Jan 1907, p. 854)

Chapter 5

Grizzly Island and the East Side 1900-1935

The Phantom Pond

When Hall McAllister was an eager teenager, Grizzly Island was a long way from anywhere. There were no bridges or ferries to it from the mainland. Not for another 40 years would it get railroad service. "The Montezuma marshes are also good sporting grounds, but they are little used because there is no way to get at them except by boat," wrote an observer in 1882. (*B & S,* 15 Jul 1882)

From Hall's home in Benicia it was a half day's sail, and to make a hunting trip worthwhile you needed to stay for at least one overnight. That meant a boat large enough to sleep on, because on the island (sometimes spelled Grisly in those days) camping opportunites were indeed grisly. If the mud didn't get you, the mosquitoes would, and there could not have been much in the way of dry firewood in those pre-eucalyptus times. It was far better to lie offshore in the west wind between the sundown and sunrise shoots.

But Grizzly was home to the Phantom Pond, a several-acre expanse of water hidden in the tules where ducks and geese would sometimes congregate in untold numbers. In 1879, at the age of 18, McAllister visited there in the yacht *John Rogers* with two companions. Having "found the pond" (triumphant underlining in his log) and shot it on the evening of November 28, the hunters returned to the yacht, then could not rediscover their prize in the next dawn, a thick fog having blown in overnight. They finally stumbled across it again when the fog lifted, and were able to get in a second evening shoot, with a total bag of 18 honkers, 1 whitefront, 1 "grey goose" (immature snow?), 12 snows, 1 swan, and 14 ducks. (McAllister scrapbook) Perhaps the same shoot was one described in an article by "Duck Call" in an 1882 edition of *Forest and Stream,* where some of the details were the same but with a greater reported kill.[1]

Putting together the two accounts and a rough map that Will Weinmann, a charter member of the Hardland, drew up for McAllister's log, the Phantom Pond can be located on the western side of Grizzly Island, probably straddling the Grizzly Duck and Montezuma clubs. In about 1980 Rich Tesene of the Grizzly Duck Club filled in what was probably the last remaining segment of its meandering, unnamed feeder slough. Though the pond was less than a half mile cross-country from the Montezuma Slough, access to it in the 1800s was tough. It meant dragging a heavy wooden boat laden with ammunition and decoys over grassland, through tules, across the feeder slough, over more high land—and then returning, even more heavily laden with decoys and game. The prospect must have discouraged all but the most dedicated hunters.

And even the Phantom had its off days: McAllister's log shows that he was skunked there once in 1880, and in 1881 his only visit produced but one lowly spoony. But Grizzly's biggest drawback was the absence of a rail connection to the City, with all of the convenience and rich-men's input that that implied.

---

[1] Hunters are hunters—after three years, some inflation of the bag was inevitable. At that, the kill was modest for those days, but there is something about a Shangri-La hunting spot that transcends the poor results of any given shoot.

Other than McAllister's, the only record we have of a hunt on Grizzly before the end of the 19th century was by one hardy pair of San Franciscans (James Maynard and Dr. E. N. Ayers), who hired a guide and his ark for a 4-day stay in 1898. Having scouted out a superb sprig pond with thousands of birds where Maynard had shot in 1897 (the Phantom?), and gotten their eye in with a "fair" bag of 70 birds on lesser ponds, the two prepared for a prime shoot on their last two days, only to be shut down by a howling southeaster that blew all the birds out of the country. Instead of hunting, they "had to console themselves with a generous stock of creature comforts [not further identified] brought up with them from the city." (*B & S,* 24 Dec 1898)

Farming the Islands and Mainland

On the islands surrounded by the waters of Honker Bay, Suisun Bay, Grizzly Bay, and Montezuma Slough, (Grizzly and its smaller neighbors, including Chipps, Hammond, Simmons, Ryer, Wheeler, and Van Sickle), farming was at first was a much more important activity than hunting. The first attempt at reclamation here was on Long Point Island (now Ryer Island) in 1865, but the effort failed.

In 1869, ranch owner Starchel Clinton Hastings drafted formal by-laws for reclamation districts, and in 1870, William L. Chaplin and three other investors had Grizzly Island surveyed and marked off into thirteen 961.63-acre lots. In 1876 California deeded the entire island to Chaplin, including an extra 3,638 acres of tideland. By that time, all but the tideland was already girded by a levee three feet high, two feet wide at its top, and ten feet broad at its base, and Chaplin was the sole owner. (Frost, p. 10)

Maps from that era shows Suisun/Grizzly Bay encroaching in a deep arc onto the island, but within a few years Warren J. Dutton began buying up Chaplin's holdings (he eventually got them all, including some taken over by the Nevada Bank of San Francisco) and expanding the acreage by building levees out into the bay. These

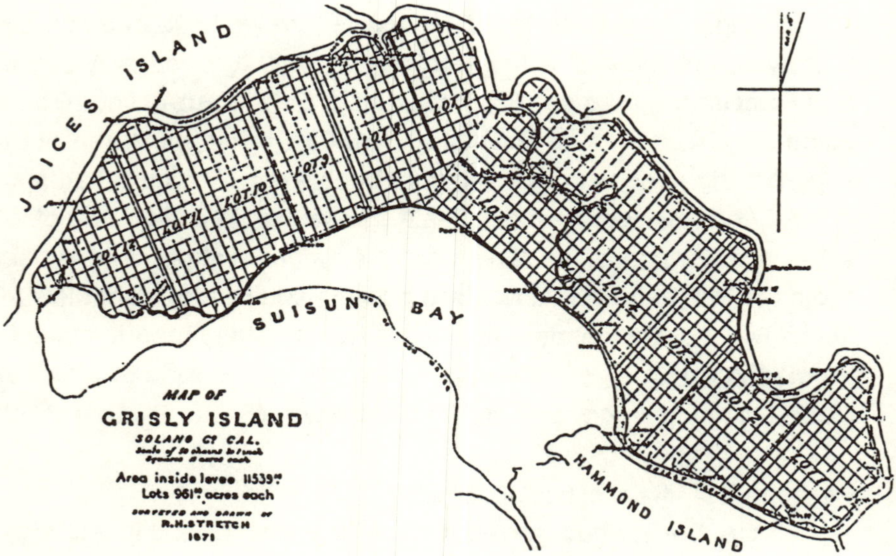

1. GRIZZLY ("GRISLY") ISLAND IN 1871

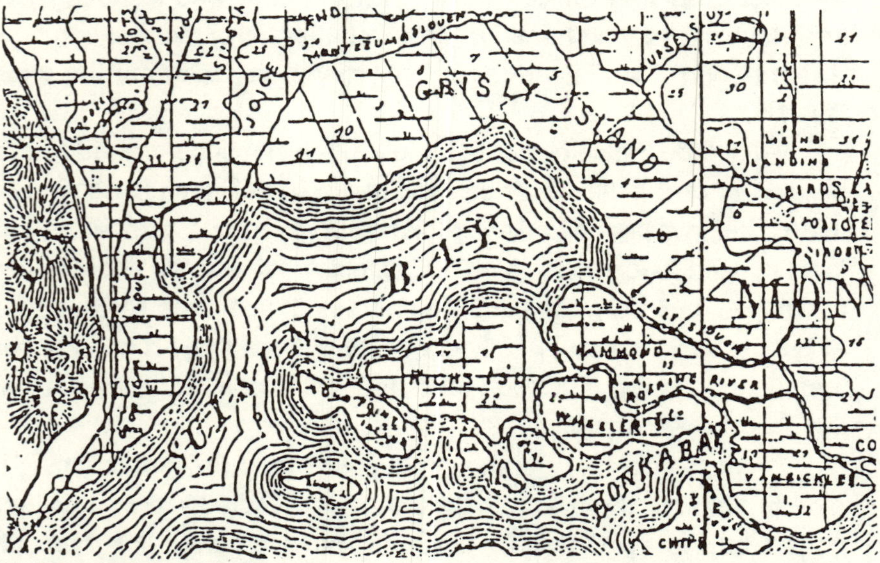

2. SUISUN MARSH EAST SIDE IN 1877

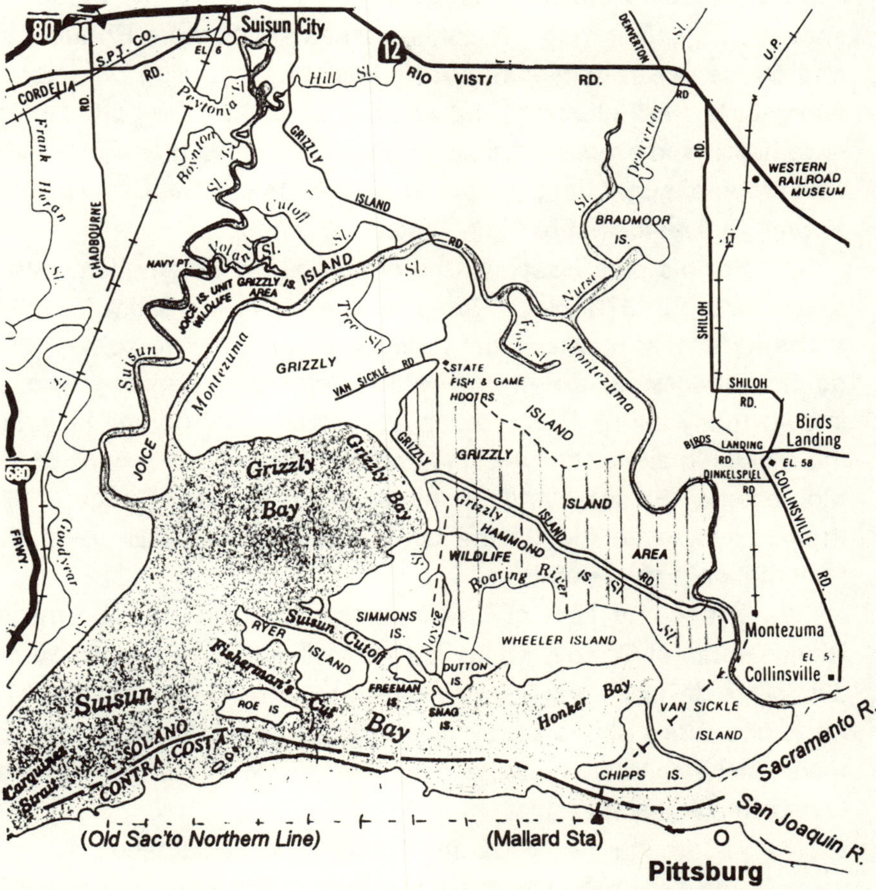

3. The East Side Today

Map 1 (facing page) shows property lines in 1871 and includes a "low water line" that equates roughly to today's leveed perimeter. **Map 2**, only six years later, indicates that some leveeing had already begun. Compare with **Map 3** above. (Cartographers are not always naturalists: not only do we have Grisly Island, but note "Honka" Bay on Map 2.) Except for the later subdivision of Suisun Bay into Suisun and Grizzly Bays and the change from Rich's Island to Simmons Island, however, most features still bear the same names, including the tongue-in-cheek "Roaring River," Joice(s) Island, and all the major sloughs.

levees eventually took in all the mudflats out to the low-water line shown on an 1871 map. Dutton's lessees at first were Portuguese and Swiss-Italian dairymen, but he later began selling property to energetic farmers interested in diversified crops. By the early 1900s, reclamation and experimental agriculture were moving ahead briskly, with crops of sugar beets, asparagus, lima beans, oats, and barley, as well as livestock of all kinds, and dairy products.

For his part, Hastings, the drafter of the reclamation district regulations, in 1873 joined Edmund Rush in laying claim to 1,481 acres northwest of Montezuma Slough and in 1875 to some 7,500 additional acres elsewhere. The first acreage is probably what is known today as the Rush Ranch, while the larger portion includes many of the marshes between the Potrero Hills and north bank of the Montezuma Slough, and perhaps most of the Potrero Hills themselves, even though the latter could scarcely fall within the reclamation district definition.

The same kind of development occurred on the smaller islands south of Grizzly, where in 1871 William S. Ryer renewed his struggle to develop Long Point and the other smaller islands. In 1873 he joined three others to lay claim to the nearly 12,000 acres of marsh that they totaled, and in June 1887 formally founded the Ryer Island Reclamation District.

All through the 1870s and 1880s, other reclamation districts were being set up on the eastern side of the marsh, including the mainland, many by men whose names became attached to islands and sloughs, or who featured in the duck hunting annals: Nurse, Stewart, Hillborn, as well as Ryer and others. The announced purpose of all of the reclamations was not hunting but agriculature, and many of the districts were split up and redrawn as land was bought and sold.

Between the Potrero Hills and Grizzly Island, just north of the Montezuma Slough, lies today's Greenhead Club, formerly part of the Hastings Ranch. According to Greenhead Club historian Marshal

Steel, in about 1880 Hastings

> dammed the Cross Slough in two places, opened a floodgate during the fresh water spring runoff, closed the gate, and now had a vast reservoir for irrigating the reclaimed land. Only a windmill was required to raise the water level in the slough sufficiently to clear the levee structure. Gravity did the rest. (Steel, p. 3)

During the 1880s and early 1890s, barges took vegetables produced on the Hastings Ranch in to market in San Francisco and Oakland, but later in the 1890s, the ranch switched to dairying, and the land reverted to pasture and marsh. In December 1895 Jack Wilson hunted with a friend on "Hastings Island" (later Bradtmoor Island and now Bradmoor Island, between Nurse and Denverton sloughs), where they "secured quite a number of mallards, canvasback, and one immense honker."(*SR,* 25 Dec 1895) Some years later, dairying on the Hastings Ranch died away entirely, the neglected levees were breached, and the land quietly reverted completely to marsh.

On the hill to the west overlooking the ranch, the beautiful Hastings manor house, carefully landscaped and tended, was the site of many elegant weekend parties. There the owner hosted his many prominent San Francisco friends, who would arrive in their finery by boat, train,and buggy. After Starchel Hastings' death, however, the house became a clubhouse for part of the Suisun Gun Club, which by then was expanding from its west-side beginnings, and after that club's dissolution, the property began to decay. Whatever heirs might have remained in California apparently had no interest in the ranch, and the sad ruins of the mansion finally burned down about 1932. "A small bronze stag was set in the center of the pond facing the entrance. It was still there in 1938 when I first visited the site, but when I checked again in 1980, some miserable bastard had stolen it." (Steel, p. 4)

Decline of Agriculture

This early agricultural development contrasts with the relatively half-hearted reclamation and farming on the west side, which focused almost entirely on grazing and duck club activity. True, Andrew Goodyear got an early start in 1880 with a reclamation district in his marshes to the south of the Chamberlain Estate, but he was the exception and never developed his land agriculturally. It was 1919 before other landowners in the west, among them G. S. Arnold, William Pierce, Fred Chadbourne, and George Tomasini, formally applied for permission to dike their lands under reclamation district rules.

Some agriculture followed, but it was small in scale and at least partly experimental. At some point asparagus must have been raised on the Arnold Ranch, for it still sprouts today on higher parts of the club, more than 75 years after it was first planted. On Morrow Island, farming persisted into the 1950s before finally dying out in favor of hunting.

Bill Richards and his wife raised "ornamental and edible plants" at Green Lodge, and William Pierce experimented with "exotic fruits" near today's Pierce Harbor, but all of these came after hunting had become the dominant activity. (For Bernice Huber, the teenager daughter of Teal's keeper in the 1920s, Pierce's melons were still memorable in 1995 as the best she ever tasted, having just the slightest tang of salt to flavor them.) As on the east side, the passing years saw the part of the marsh that had been reclaimed for agriculture go back to pure duck hunting. A topographic map, for example, indicates that in 1949 the Sunrise Club was drained farmland, but it has been nothing but a duck club for at least the past 40 years.

On Grizzly and its neighboring islands, agriculture at first flourished but eventually withered. Why the decline? It was not absence of ingenuity or energy or precedent. Disastrous floods in 1896, 1902, 1904, and 1907, for example, failed to drown the farmers' early enthusiasm, nor did later inundations.[2] The reason was not infertility, for the land in the early 1900s was extremely productive for the times: 70 sacks of barley and 30 sacks of beans per acre without any irrigation on Ryer Island, for example, which also boasted a hemp mill. Later, asparagus, sugar beets, alfalfa, vegetables, and wheat followed, supporting an agricultural workforce of perhaps 1,500 on that island alone, which even had its own elementary school.

Flooding and diking meant dogged labor in the face of adversity, but the ranchers were inventive as well as hardworking. In November 1902, a 12-mile phone line was established linking the Arlington Hotel in Suisun with the Barnhart Ranch two miles to the east and thence via barbed wire property fence lines to the Lemmon, Robbins, Hastings, McCarron, Lewis Morrill, and Rush ranches. The reception over the barbed wire was said to be so excellent that the day the line was opened, the Hastings keeper, Joseph Hoyt, was kept up nearly to midnight answering calls from his friends in town. (*SR,* 21 Nov 1902)

With the kind of get-up-and-go that can survive regular floods and turn a fence line into a phone line, the farmers on the marsh seemed to have a bright future, but through no fault of their own it was not to be. The culprit—if it can be called that—for the decline of agriculture was the saltwater intrusion that poisoned the soil as the fresh, upstream waters of the Sacramento and San Joaquin rivers were drawn off for other human needs. By 1976, the fresh-water flow in the marsh had dropped to half of its turn-of-the-century level, but agriculture was in decline long before the 1970s.

The farmers may have been the losers, but the winners in the long run were the marsh itself and everything wild that lived there.

---

[2] In 1938, according to one old-timer, "you stand on Ryer Island and look west and all you can see is water. It looked like the coast all the way to the hills almost." (Craven)

Hunting the Northern Mainland

By 1895, with Grizzly still inaccessible and focused on farming, Suisun hunters began responding to hunting pressure in the west by pushing eastward on the mainland both north and south of the Potrero Hills. The Vest place, just north of Hill Slough and owned by an absentee Klondike millionaire, and the adjoining Frisbie and Lemmon properties were an early duck-hunting mecca in that pre-automobile era. (The marshes remained in private hands up until the 1970s, when they became the Hill Slough Fish and Game Wildlife Management Area.)

The hunters either worked the sloughs, shot in unsupervised marshes like Vest's, or went out to specific ponds owned by local farmers and often in their company. As the marshes just east of Suisun came under pressure, hunters pushed still further east, to Denverton and beyond. The newfangled automobile began to give hunters extra range, and before long the marshes east of the Potrero Hills were in easy reach of Suisun. The major landowners there were the Stewart, Morrill, and Lang families, with the Stewarts having laid claim (by name anyway) to the best hunting ponds.

The earliest named club was the Potrero Gun Club, launched in October, 1899, and located on the "Lang and Morrill marshes between Denverton and Nurse Sloughs." Unlike the shacks that passed for clubhouses elsewhere, the headquarters of this group was noteworthy for its luxury as the former residence of Fred Morrill. The implication is that the homestead had been given up for some reason, possibly in connection with the discontinuation of farming on the property.(*SR,* 6 Oct 1899) Perhaps the fancy accommodations were the drawing card that Claire A. Morrill, who had been shooting the area with San Francisco guests at least two years earlier, needed to entice five city hunters into forming the club. A seventh member, Max Kerchaw from New York, joined them in November that year and gained instant notoriety by "maiming several decoys," but retrieved his good name by helping gather 50 pounds of mushrooms in the hills overlooking the marsh. (*B & S,* 25 Nov 1899)

There is still a Potrero Club in the northeast corner of the Suisun Marsh, and it doubtless includes at least part of the original, though it lies to the west of Nurse Slough. Next door is another club formed in 1899 but with no name attached, almost surely the one known now as Pickleweed Ponds, whose present-day owner is a descendant of Frank Branscomb, one of the seven founders of the unnamed club.[3]

"Stewart Ponds" was a generic term used for the whole area. Almost certainly, two of the larger ponds are those shown on a 1974 map as lying on the eastern and western borders of the Pickleweed Ponds club. North of there and now sliced through by Highway 12, was another "Stewart Pond" that used to be a canvasback haven. There were undoubtedly many others. In 1902, Joe Harvey (later of the Dinkelspiel syndicate that bought the Chamberlain tract) arranged with other "city capitalists" to lease some of them for the 1903 season. This included only part of the wetlands, however. Stewart ponds reportedly were part of at least three other clubs formed in the area over the next few years, to say nothing of being favorite spots for various Suisun and San Francisco freelance hunters. The property owners included the Morrills, one of whom formed his own club (the King Shooting Club) in 1903, and another of whom rented out some of the "Stewart" ponds on his property to a newly formed Redhead Club in 1905. (*SR,* 21 Nov 1902, 16 Oct 1903, 17 Nov 1905)

---

[3] This lady, Jean Davisson, like all members of her family an ardent duck hunter, during an active dawn shoot in the 1950s suffered an unusual gun malfunction; she became so excited as thousands of birds worked in that she mistakenly included her lipstick with the fresh shells she was shoving into her gun's magazine, where it provided one of the most thoroughgoing and colorful jams ever suffered by a Model 1100. (Davisson)

Regardless of who owned what, the ponds were a hot hunting commodity. In 1907 the four members of the Stewart Club, including Fish and Game Commissioner Lendal M. Gray, got 80 birds on the opener, one of the best records on the whole marsh that day. To open the 1909 season, Thomas T. C. Gregory, the prominent Suisun lawyer who had been such a thorn in the side of the Cordelia Club in the 1890s for his defense of trespassers, hosted the US Navy's Admiral Holiday, who, all the way from Annapolis, already knew enough about Stewart pond shooting that he had craftily arranged to inspect local naval facilities just in time for this hunt. On a November shoot that same year, Frank Stewart and three friends limited out with 25 birds each by 7:30 in the morning, and Frank repeated the feat in 1911. (*SR,* 4 Oct 1907, 8 Oct 1909, 20 Oct 1911)

The Stewart Club is what is now called the Duck & R Club on the SRCD list, but one of the owners is Wes Stewart, a direct descendant of the founder. Like Pickleweed Ponds, the Duck & R has been passed down within the same family for nearly a century.[4]

Claire Morrill, having moved away from the family homestead in about 1901, reestablished himself on the southern flank of the Potrero Hills. He forsook the club philosophy in favor of having one-at-a-time guests who shot a single pond within sight of the owner's house. Claire went in for careful management and strict observance of the game laws, never violating the two-shooting-days-a-week custom of the times. By 1907, when many clubs were suffering seriously from poaching on non-shooting days, he was able to ensure

---

[4] In 1910, the Stewart Club consisted of five members, two of whom were William Denman and William Kent. Later, in 1912, Denman became the law partner of G. S. Arnold and with him bought the West-side acreage that became the Arnold Ranch. Did Denman introduce Kent to Arnold? We don't know, but Kent was father of Elizabeth Kent, in 1915 to become Mrs. Arnold, my mother. The Suisun Marsh may have forged the Kent-Arnold link that's responsible for my being here.

that birds on his pond were left in peace most of the time. An excellent caller and an excellent shot, he stood out from the neophytes who were common among the new club owners of those days. As a result of his practices, he and anyone lucky enough to be his guest turned in consistently high kills. (*SR,* 6 Dec 1907)

The Southeast Mainland

At about the time the Suisun shooters were pushing in from the west along the north flank of the Potrero Hills, other hunters were moving in from the south and east. In 1902, three members of the Black Jack Club pulled up stakes from Sherman Island along the San Joaquin River and, taking the club name with them, moved in close to Collinsville. The next year saw the founding next door of the 600-acre Montezuma Club, which included W. W. Richards as well as Black Jack transplants Bert Wyman and William ("Hardmouth") Swain (did he do his own retrieving?), Joe Rogers, Bert C. Scott, and one or two others. The founders put $2,000 into the improvements, which included a handsome, two-storey clubhouse with a view over the marsh. (*B & S,* 4 Oct 1902, 3 Oct 1903; *SR,* 16 Oct 1903) Access to these clubs was possible by land from Suisun or, from the East Bay, by boat across the Sacramento River from Antioch. (Travel in those parts could be dangerous. A broken arm, multiple contusions, and internal injuries to the clubs' members followed a 1905 wagon accident one dark night, and in 1906 two Montezuma Club hunters barely escaped drowning when a howling norther drove their flimsy "pickax" boats into the tules far from the clubhouse.) (*B & S,* 7 Jan 1905 and 6 Jan 1906) A Gadwall Club, owned by medical doctors, existed near Collinsville in 1909, and a Youemace Club was set up in the same area about 1910. (*B & S,* 26 Dec 1909 and 20 Dec 1913)

From Meins Landing in the south to Tooby Farms in the north, duck clubs still hold sway on the east side, but the passage of time has silted over much of their history. The Black Jack, Gadwall, and Youemace clubs have all vanished, their marshes now owned by Fish and Game. Only the Montezuma can be traced to a present-day club, but it has been relocated miles to the west for more than 80 years.[5] The waters that were christened "Stewart ponds" doubtless still exist— you can see a number of good candidates looking west from Shiloh Road—and ducks almost surely follow their distant ancestors' habit of using them, but the old property lines can only be guessed. Probably the most surefire way of finding a former Stewart pond now is to seek out the most productive blinds in the area.(*B & S,* 8 Oct 1910, 20 Dec 1913)

Early Hunting on Grizzly and Neighboring Islands

On the 1901 opening day, Suisunites Jack Wilson and his perennial hunting companion Jack Harper went to some unknown location on Grizzly Island. The two Jacks were forever trying out new areas, and this time they returned with a reported bag of 22 ducks between them. This was a poor score for the enterprising and usually successful pair, and there is no indication that they ever returned, but Grizzly Island's hunting days were coming. (*SR,* 4 Oct 1901)

After the slow start by Wilson and Harper, other locals eventually began going to Grizzly, at first alone or in pairs, but by 1907 in larger groups. The island was already gaining a reputation farther afield as well. A May 1908 article in *Grizzly Bear* concedes that the Suisun Marsh as a whole has "many suitable nooks...but

---

[5] For unknown reasons, the Montezuma Gun Club picked up stakes and transplanted itself from the Collinsville area first to some unknown point on Joice Island (1914-1915) and then to the west side of Grizzly (looking for the Phantom Pond?), some time thereafter. The link of continuity is Bert Scott, who was a founding member in 1903, president when the club was incorporated in 1924, and still in charge when it dissolved as a corporation in 1944. May his ghost still give the Montezuma a friendly haunt.

Grizzly Island is the favorite of a great many [hunters]." In 1909, Howard Vennink, who had arrived on the island in 1906 and quickly established a reputation as an experimental asparagus farmer, pioneered the idea of leasing hunting rights in the agricultural off-season. His first property was Tract 10, and his original house is now the clubhouse for the Payton Place Club. Other clubs now located on Tract 10 are the Four Winds, Grizz/Fizz, West Wind Sports, Garben Ranch, Wild Turkey, most of the Honker and Mendoza, and parts of Tree Slough Farms and the Long Point West clubs. Later, Vennink is supposed to have acquired Tract 6 property, which today is the location of many clubs (Sprigsville Ranch, Sleeping Pintail, Marsh, Bulrush Farms, Gang Bang, Windmill, and Horseshoe), plus a small part of the Grizzly Island Game Management Area.

Two different groups of Suisun hunters snapped up Vennink's 1909 lease offer, but their clubs' names (if they had any) and exact locations are unknown. Other farmers were relatively slow to follow suit, but with the passing decades and the degradation of the land's fertility, fish and game gradually became the main—and finally virtually the only—human activity on the island.

The Southeast-side Island Clubs

In December, 1910, Warren Dutton sold all 5,500 acres of Wheeler, Hammond, and Rich Islands to agricultural developers, having already divested himself of six other parcels, ranging in size from 500 to 2,000 acres to six different farming hopefuls. "All of these tracts not now under cultivation will be reclaimed. This change will see the passing of quite a bit of duck shooting ground," went one gloomy report at the time. (*B & S*, 10 Dec 1910)

There was reason for the gloom. In 1908, the opener on Ryer and Wheeler islands had seen limits of sprig and mallard for a half dozen hunters that were impressive enough to merit commendation in the *Breeder and Sportsman*, but now hunting seemed doomed. (*B & S,* 10 Oct 1908) Despite Dutton's sale and the agricultural plans, however, clubs were soon established on the same marshes supposedly slated for draining and reclamation. The first was the Chipps Island Gun Club, whose nine members opened the 1910 season with a "big mallard shoot," and which may have included the island's entire 975 acres, now divided among the Fin and Feathers, Dante Farms, and a third, unnamed club. The original club was incorporated in 1924 by a different group of ten hunters but under the same name. The Ryer Island Gun Club, launched before 1908 by San Franciscans, was taken over in 1911 by local hunters (C. M. Jones, Jack Perkins, P. Porter, and others), and today belongs to the Catellus Development Corporation. The Wheeler Island Gun Club got going in 1912 with ten members, mostly from San Francisco. Finally, 1914 saw the establishment of the Chiquita Duck Club (A. G. Hogben, C. Asselena, Dr. Chismore, A. W. Sexton, and others) on Hammond Island, on what is now Fish and Game property. (*B & S,* 8 Oct 1910, 3 Feb 1912, 30 Nov 1912; 7 Nov 1914; CR)

Long after the west side marshes had all been sold or leased to clubs, there was still room for freelancing on the east side. Some time before 1920, the late Bill Agler, a Concord resident, used to hunt Van Sickle Island by taking a rowboat across Honker Bay, then walking two miles to the hunting grounds. Bill and a companion would tote all of their gear and supplies for two or three days across that unforgiving marsh, set up camp, and shoot until they ran out of everything but ducks, then trudge back to the bay and row home. He later helped found the Concord Club in the same area, and though his name does not appear in the articles of incorporation, he is prominently displayed in early photographs of the club, usually festooned in sprig. He was probably one of those types who paid his dues more in hard physical work than in cash or paperwork. His later protégé was Les Jacobson, one of the senior members of today's Concord Club, who still remembers him with affection and some awe.

Although the east-side clubs doubtless suffered from the same poor seasons that afflicted the west side in the middle 'teens of this century, the ducks came back in the 1920s, and so did the hunters. After the Concord Gun Club became incorporated in December 1923, a whole rash of incorporations on the east side followed: Montezuma Gun and Wheeler Island clubs in 1924 (the latter changed hands that year), and in 1925 the Delta Sporting, Rich Island Gun, and Sprig Farm clubs. These have proven to be remarkably durable institutions which, except for the Delta Sporting (was it a forerunner of the Delta Farms or Delta King?), have retained their names down to the present day. (CR)

The Coming of the East-Side Railroad

As was the case in the west, the blossoming of clubs in the east was largely thanks to the coming of the railroad, and it is not surprising that west-side duck hunters with money to invest in a new railroad took a leading role in plotting where the tracks were laid. As early as 1892, the *Rio Vista News* reported plans for a line running from San Francisco via Pittsburg, Chipps Island, Van Sickle Island, across Montezuma Slough and so on to Sacramento. Nothing more came of this, but in 1906, Joe Harvey (Harvey Club) joined A. L. Schindler and Thomas T. C. Gregory (both of the Sheldrake Club), Louis Sloss (Family Club), Henry Klevesahl (Seymour Club), William Pierce (Suisun Gun Club), and several other notables from Suisun and San Francisco to launch the Vallejo and Northern Railroad Company, which, after several false starts and renamings (and abandonment of Vallejo as a terminal), became the Sacramento Northern, "the longest interurban line in the United States and perhaps the world." Like the Southern Pacific, the line ran east from Oakland, but instead of turning north at Martinez, it followed the planned 1892 route. When completed, it ran 183 miles from Oakland via Sacramento to Chico and Redding. (Swett, p. 3)

Regular service started on this line in August 1913 (note the coincident founding of some of the southeast side clubs) but was suspended when disastrous rains hit the delta in January 1914, putting both Southern Pacific and its new rival out of action and forcing the latter to put in 8,000 feet of new trestle. In comments reminiscent of the SP's experience, reporters noted that the track was "on marshy land, where the ground is constantly sinking." But at least the Sacramento Northern's trolleys were lighter than the transcontinental trains, and trestles could usually serve instead of the massive roadbeds needed by the SP.[6] (Ibid, p. 102)

It was still a shaky start in more ways than one, however, and it was only some years later that the line and the duck clubs it serviced really established themselves. By the 1920s, the hunters were already exploiting their new access, not always to the railroad's benefit. "Even a crack limited train could be stopped at an obscure depot if requested ahead of time. This was particularly true north of the river [i.e. the Sacramento] where virtually all of the local passengers consisted of duck hunters. One such hunter stopped all trains for one day when he missed his target and instead severed a trolley wire." (Ibid, p. 101) It was fortunate indeed that many of the railroad's backers were sympathetic to hunting, or the whistle stops might have been removed after this incident.

The corporate sympathy was, however, based on more than just duck hunter camaraderie. The line needed full-fare passengers, and by the end of the 1920s it was making a special effort to cater to sportsmen by billboard advertising, running specials for hunting groups, and even using its steam launch, the *Countess,* to help clubs not serviced directly by the rails: Rich Island Gun Club, Sprig Farm Association, Wheeler Island Gun Club, and St. Germaine Duck Club. The schedule for the *Countess* involved connections with two east-bound trains for delivery of hunters to the clubs on Tuesday and Saturday afternoons and connections with two west-bound trains on

---

[6] Or so engineers assumed. In 1951, when the railroad tried to service a steel mill in Pittsburg, a whole section of the Yolo causeway trestle collapsed under a heavy train. (Nadeau)

Sunday and Wednesday evenings. (Note that this confirms that the two-shooting-day week was still being observed at that time.)

The *Countess* also had special runs on a four-day weekend at the end of September 1929, perhaps for some club socializing. (It is hard to imagine that there was such a thing as today's "work weekends" for those gentlemen of leisure.) The Sacramento Northern, like the SP, also made an effort to protect its clients' privacy: an internal corporate letter in September 1929 specified that "under no circumstances will the *Countess* be permitted to take poachers to any property of the four clubs or to any of the adjoining islands." (SN Letter)

For clubs closer to the main line, the railway also had special stops for shooting days and the evenings preceding them. Some of these were stations named on the map (Mallard, Chipps, Spoonbill, Dutton), some do not show on the railroad's map (Honker, Dutton's Ferry Road, Millar), and one is only identified as "the first road crossing west of Lisbon Trestle." (SN Memo 12 Sep 1929)

Why the west-side investors ventured into the shark-infested waters of railroad competition remains unclear. A common motivation in those years was profiteering. One first bought land, then ran a railroad into or past it to improve property values, and finally made the decision to hold or sell based on an assessment of the railroad's probable fortunes. With the clear example of the Dinkelspiel syndicate's success in boosting marsh values in the west, there should have been some effort by the investors to buy property near the new line, but that apparently did not happen. Maybe it was just the altruistic desire to promote hunting on the marsh.

Wheeler Island Club

Of the clubs on the southeast side, the most documented is Wheeler Island. In 1990 Ron Dysart published its history back to its incorporation in 1924, but in fact the origins of the club go back to 1912 and possibly further. In 1908, the Olympic Club's Al Hampton and two other hunters (Billy Corbett and Charles Hern) got limits of mallards on the "Parkinson Place" on Wheeler Island, and four years later A. W. ("Cap") Simonton and nine other San Franciscans opened a formal club there. Simonton later went east to take charge of the DuPont factory's traffic office in Delaware but returned to California for the hunting whenever he could. The Wheeler Island Gun Club later became the Wheeler Island Land Company, then the Black Mallard Duck Club, and finally reverted to the Wheeler Island Club. Originally 1,054 acres in size (now 971 acres), Wheeler Island still had the traditional ten members from its formation to its incorporation in 1924, but its membership quickly expanded to 17 and then, in 1974, to the 20 originally anticipated. Dysart's history chronicles not only the members and their various projects but also a good many anecdotes and observations.

✧

The Good Old Days

Billboard ad for the Oakland Antioch & Eastern RR Co. (forerunner of the Sacramento Northern) dating from 1917

Chapter 6

20th Century Survival Fights

During the first half of the 20th century, the threats to the marsh were less intense than during the second half, though some warning signs can be found all the way back to the first decade. Happily, the economic booms and busts hit everyone alike and did not result in wholesale property transfers. The good times (which also, fortunately, seemed to parallel the years of best duck migrations) made club owners want to hold onto their properties; the bad times just meant that there were no buyers.

By the end of 1910, the Dinkelspiel syndicate had sold off for duck clubs almost all of their 1905 purchase, reaping a threefold profit from their investment. In a recurrence of the 1905 rumor, the last known sale (of 1,000 acres, to the Presley Company, which eventually became the Sheldrake Club) was billed as a prelude to property's subdivision into small holdings for farmers. If this had happened, it would have driven a wedge of farmland south from Suisun between Joice Island and the west-end clubs.

As if reporting a done deal, the *Breeder and Sportsman* wrote "(the property) would have made one of the best preserves on the marsh," but, it added, if farmers came in, shooting could be ruined for everyone in the area. The surrounding duck club owners might be forced to "join the reclamation movement or sell out." They would have little choice, for the soil "is about as rich as can be seen anywhere and will raise fruit, berries, or vegetables in luxuriant crops."[1] (*B & S*, 19 Jan 1910) Fortunately, it didn't happen.

During the meager duck flights preceding and during World War I, there is no evidence of fights to preserve the marsh, probably because there were no believable plans to do anything else with it. The drought years of 1920, 1924, and 1926, and again in the 1930s, did lead to salt-water intrusions into the delta and renewal of plans to freshen the water serving Grizzly Island farms, but these all foundered for various reasons (Chapter 1). A possibly key role in blunting enthusiasm for development was a Division of Water Resources bulletin which stated that the marsh was more valuable for duck clubs than for anything else. (DWR-28, 1931, p. 35)

The Depression and Aftermath

The stock market crash of 1929 heralded not only the 1930s Depression but devastating droughts across the nation. It was a bad time for humans and worse for ducks. Hoboes, known as "knights of the road" in those days, rode the freights past our club, and those who walked the tracks would often stop to ask for work or a handout. Big Stan insisted that no one be turned away, but Gus Davis, our crusty keeper, objected to parting with good food, and when Big Stan wasn't looking he would sometimes substitute warmed canned dogfood—or sometimes just a cold can of it, the label removed. It was a sign of the times that even such humble fare was accepted with philosophic resignation if not wild enthusiasm.

For the ducks, there was no substitute for the water they had

---

[1]This exaggeration may have been merely a gimmick to raise the price of the remaining marshland, which, though untested for agriculture, had been tried and found less desirable for hunting than the earlier club properties.

to have in their nesting areas. When that dried up, so did the waterfowl hatches, and the majestic fall migrations shrank to a relative trickle. The limit, which had held steady at 25 for more than 20 years, dropped from 15 to 12 to 10, and the season shortened by two thirds. A possession limit was imposed in 1931, and by 1935 it was only equal to, not double the daily bag limit (Appendix C).

The previous two decades had seen the rapid development and expansion of clubs, despite the periodic drought years from 1917 onward. But now the duck hunters could no more escape the times than could the homeless and ducks. The Roos brothers first stopped entertaining at their luxurious club and then stopped coming altogether. The echoing, ghostly white building with its stained glass windows and colonnaded front porch, no longer attended by a keeper, stood empty on the marsh, its furnishings removed.[2] Remarkably, however, the old landmark survived the Depression and World War II, only burning to the ground sometime in the 1950s, when destructive vandalism again became fashionable for the have-nots. (Maybe during the Depression a roof—any roof—was too valuable to destroy deliberately if there was no other close by.)

Through it all, the club owners clung to their wetlands. Granted, they probably could not have sold even if they had wanted to, because the few people still rich enough to afford a club had little incentive to risk their remaining dollars on such luxuries. But there was no sign that the club owners were willing to forfeit their beloved

---

[2] Toward the end of the 1930s, Big Stan leased the Roos club, and my oldest brothers sometimes hunted it. They would overnight there on the floor, equipped with nothing more than their sleeping bags, guns, and a Baby Ben to wake up by. (How many amiable ghosts must have looked down on them!) Breakfastless, they would row two little one-man skiffs down a long, narrow channel to the main pond, probably the original Sixth Reach Pond of the 1800s.

marshes for taxes or try to sell out at a loss. Their fiercely possessive love remained as strong as ever.

If the Depression made duck clubs just another burden for some owners, for at least one it spelled salvation. Clifford Wagenet, who had bought his club on the east side near Denverton from the proceeds of a successful wholesale flower business in the East Bay, found that his love of hunting helped him through difficult times. When personal and professional adversities combined with the Depression to force the closure of his business, he quietly retreated to the marsh with his family and from that lonely outpost rebuilt his life. (There are far worse places for such a project!) Wagenet's descendants still own the property.

As the country made its slow economic recovery at the end of the 1930s, duck hunting picked up. Although it was still far from a carefree time, the marsh itself was not under attack, and it became a refuge from the other problems of the world. Just the recollections of getting there are an exercise in nostalgia for some. Pete Arnold jotted down his memories in 1982:

> I made the trip by rail once or twice, taking the train from Kentfield to Sausalito, the boat to the San Francisco Ferry Building, and another, larger ferry to Oakland. From there a three-car "rattler" provided local service to Sacramento. The cars must have been the oldest on the line, the woodwork needing varnish and the plush of the seats suffering an acute case of mange. The conductor had been on the run so long that when he punched my ticket he once said, "Teal, eh? You must be an Arnold. I used to go bass fishing with duck gizzards at your dad's place." The ticket from San Francisco, as I recall, was about $1.90. [Author note: In 1896 it was $2.50 for a round trip.] It was a good half day's trip from Kentfield to Teal by rail and ferry.

Teal, like the other stations between Benicia and Suisun, was only a whistle stop, so the trains had to be signaled. Recalls Stan Arnold, "I can't think of anything calculated to make a boy feel more important than to stand out on the main transcontinental train tracks with a little red lantern, and flag down a great huffing monster. It was about as impressive as stopping the sun or the moon."

By car—ours was a tiny, grey, parrot-nosed Willys—the trip was faster but probably twice as long as it is today. In those days if we drove up on a Saturday morning we could listen to an Ivy League football game on the radio (shows you how long ago *that* was—out here they didn't even *broadcast* those pipsqueak Big Ten games in the 30s). The Black Point and Sears Point cut-offs (today's Route 37) were still two-lane private toll roads, and on the east end the road wound south of American Canyon before joining Route 40, today's Interstate 80. As Pete remembers it,

> We came off on Abernathy Road, through pear orchards and dairy cattle pastures that eventually gave way to more alkaline soils supporting little more than salt grass. ...The road climbed onto the levee of Wells Slough, following it to the railroad crossing..., How treacherous that levee road was in the rain....(Once) Gus the caretaker had to come out with the tractor and tow us. [Author's note: Amidst all the other changes, levee roads are still the same.]

World War II

With the start of World War II, the Arnold family's connection with the marsh was severed. Big Stan's death resulted in the immediate sale of the club, the older brothers went off to war, and my hunting became strictly local. From 1941 until my return home in 1979, Suisun was only a wistful memory. For those who still hunted there, gas rationing (four gallons a week) and ammunition shortages were daunting challenges. Still, most of the clubs stayed active during the war. People understood about priorities in those days.[3]

Typical of the breed was Bill Coon, at that time a young

---

[3] In 1944 I somehow got a whole case of 16's —and literally counted all 500 of them, like a miser with his gold. I commuted to the Greenbrae marshes by bike, once putting a dent in the barrel of my L C Smith (still in service) by catching it in the spokes. Rubber boots were unavailable, so I just got wet—but I was used to it. All that for maybe a couple of goldeneyes or ruddies.

dentist just out of college, and later one of the marsh's most effective defenders. He recalls having been a guest at the Sheldrake Club one foggy December morning and having a spoony slip out of the mist, catching him by surprise. It might have become his first duck kill ever, but he shot twice and missed, and the bird vanished in the murk. Oh yes, the other memorable thing about that day, he remembered, was the Japanese attack on Pearl Harbor.

For all the war's upheavals, which fortunately did not involve any military use of the marsh, the clubs survived, their owners still men of considerable power. Discovered in October 1994 in the bottom of my socks drawer at the Teal Club was the faded carbon of a Southern Pacific notice titled TRAIN SCHEDULES FOR DUCK HUNTERS —OCTOBER 14th, 1944 to JANUARY 1, 1945, Incl. It originated in the Office of the Vice President in charge of System Passenger Traffic, and it shows two special trains in each direction to carry hunters to and from their west-side duck clubs for Wednesday, Saturday, and Sunday shoots. Outwardly unimpressive, the crudely typed paper revealed the passengers' clout. Remember that this was before the age of interstates and air freight, and rails were the only way of getting anything across the continent in a hurry. Remember, too, that this was the last full year of World War II, and every tank, bullet, and man needed to finish things off in the Pacific (and a good deal going the other way for the unfinished war in Europe) had to pass this way. The railroad was a giant artery through which the life blood of the war had to pass in an unbroken stream, pausing for nothing...

Except the duck hunters' three-car rattlers on shooting days.

Early Postwar Period

During and after the war farming and ranching on the marsh continued to decline. Over on the east side, the big Baby Beef Company, founded in 1927, and the Lawler/Vaughan cattle-feeding operations dating from 1936, gave way to hunting. The late Bob

Rogers of Fairfield rented the north side of Bradmoor Island, formerly called Hastings Island, from the Lawlers. On the east side of Bradmoor was the Poopdeck Club. These marshes are now being shared by the Wild Wing, Flying D, and Overlook clubs.

In the late 1940s the various Montezuma Slough docks for taking produce to market were being abandoned one by one. The potato farming that had once helped keep Meins Landing in business as a port died out at this time (the wetlands that bear the same name now provide some of the best hunting on the marsh), and other docks—Duttons Landing, Beldons Landing, and others—followed suit. The end of grazing was not far behind. In 1947 a milestone was passed when the Avila family sold the last dairy farm on Grizzly Island. The beef feeding operations became history after John Lawler died in 1961 and Bud Vaughan retired in 1965. Glen Lawler at first leased out blinds while pasturing his cattle in the area but gradually sold out all his property to duck clubs. Agriculture as a significant part of human activity on the marsh was over.

The Sacramento Northern, which gave up regularly scheduled passenger trains in 1940, also appeared to be circling the drain in the early postwar years, but to the end it played an important role for the east side clubs, still providing service, if no longer regularly scheduled. John Ward, a retired hunter who used to commute to the marshes courtesy of the line, tells of the standard Sunday afternoon ritual at his club, which was located some distance up the Montezuma Slough from the train stop. The problem for club members lay in their reluctance to leave the clubhouse and its convivial drinking in time to meet the train. Only when they actually spotted the trolley would there be a mad scramble to pile men, guns, dogs, gear, and game into their one-lung launch and head for the station. It was no contest even on a favorable tide—the train always won handily—but the launch would hoist its flag on a mast specially rigged for that purpose to tell the train's engineer they were coming. On arrival at the station wharf, they would find the train crew puzzling over a hotbox or some other mysterious malfunction that had caused a breakdown, a problem that instantly resolved itself as soon as the hunters had flung their sacks of ducks—including those for the train crew—into the baggage car and piled aboard, pulling out hip flasks to celebrate their arrival.

The railroad was a facilitator in other ways as well. In those days few clubs had telephones, and a trip to the marsh was one way staying out of touch. At least one San Francisco hunter turned that fact to an advantage that had nothing to do with hunting. Having established a close relationship with a Berkeley belle, he would bid farewell to his wife on Friday evening, board the Sacramento Northern's duck hunter commute train in Oakland with his fellow club members—and get off two stops later. His companions would continue on, shoot both their and his limits over the weekend, and, by prior agreement, meet him on the same train's return trip on Sunday night. The arrangement suited everyone, and all would arrive at the end of the line together, weary and happy from their respective exertions. How he managed things in the off season is unknown. (Perhaps his family never knew there *was* an off season.)

The Upbeat 1950s and 1960s

As the skimpy flights of the late 1940s gave way to an abundance of ducks in the 1950s, hunting became more popular. A 1959 survey showed that there were 110 duck clubs comprising 28,615 acres of marsh, slightly more than the acreage reported in 1931. This was 60 percent of the Suisun area protected by levee, and a state government document of the time came to the same conclusion that other researchers had reached in 1931: that the economically most profitable use of the marsh was for hunting. If this caused glasses to clink in the clubhouses, however, the celebrations were premature.

Nevertheless, for the time being, Suisun Marsh hunting not only improved, but it also became available to many people who in earlier times could not have afforded it. The Grizzly Island Game Management Area opened in 1948 when limits were still relatively low, and, as restrictions were relaxed, it became a mecca for independent hunters without the means to join a club. Far from being a true refuge, the management area's main goal was simply to lure waterfowl away from the Central Valley, where they were playing merry hell with the grain and rice harvests, and get them in range of Bay Area gunners. The ducks and geese flocked in from central California, and the hunters flocked in from the coastal towns and cities to greet them.

At that time, however, the only way of getting on Grizzly Island by car from the mainland, a distance of perhaps 150 yards across the Montezuma Slough, was on a single, small, slow ferry. This faithful craft had to toil six to eight hours to handle the incoming hunters' cars before a shoot day—and just as long to get them off afterwards. Island residents had to have special passes to let them jump the queue. Talk of building a bridge to replace the ferry was heard as early as 1950, but a full ten years would pass before the island's famous asparagus queen, Miss Louise Kellogg (no known relationship to the Cordelia Club Kellogg), cut the ribbon for enthusiastic motorists in December 1960.[4]

Despite the transportation bottleneck , farmers and cattlemen with land near the management area had already seen the potential and begun leasing out blinds. From 1952 to 1954 my old friend Mat

---

[4]George Carnegie, the one-eyed Scot who was keeper at the Pintail Ranch Club and moonlighted as ferryman during shooting days, would look for his club's communal car to join the queue. When it did, he would come running, shouting, "This way, Doctor, this way!" and would conduct it to the head of the line. Nobody seemed to notice that the wet, unshaven, red-eyed hunters and their retrievers in the "doctor's" car did not exactly fit the part. (Keeler)

Keller hunted with his father and brother in one such operation. They would come up the night before the shooting day in a 4-wheel-drive World War II surplus ambulance, cross Montezuma Slough before the dawn rush of hunters was on, and sleep in their 12-foot trailer at a commercial club south of the refuge, where they had two double-barreled blinds. There was a clubhouse, but they never went to it, Keller senior apparently feeling that their fellow hunters' society might be a corrupting influence on his sons.

All across the marsh it was a time of change, and not everyone approved:

> In the late nineteen-forties and into the fifties...a newly affluent hunter bought into the marsh. Many new and smaller clubs began to form. Around the perimeter of the marsh...trailers, quonset huts, and shacks went up to accommodate the new hunter, grimly set on that full limit of sprig. A few, very few, new and excellent clubs were developed. On the old established clubs the hunting became spotty and the hunters became peevish. (SSCD 1, 1966, p. C-11)

But the old also endured, at least for a while. Melvin Frohrib, who as a young man used to get free shooting on an unused part of the Teal in exchange for field work, remembers the style of the club under President Carlisle:

> Mr. Carlisle's guests would get off the train at Teal Station in their business suits, go down the boardwalk and through the gate to the front porch. They would be shown into the main room, where an immense floral display was always on the dining table. A butler served cocktails, and servants waited on table for a catered supper with crystal and silver. After dinner there were high-stakes poker and domino games. The following morning, the guests found waiting for them everything needed for the day's shoot, from waders to hunting jackets to shotguns and ammunition. (Frohrib interview)

But Mr. Carlisle's time was running out. When terminally ill he still would come up to the club, ringing across from his bedroom to the keeper's house whenever he needed assistance. After he died in 1954, his son kept the club but without quite the same style. (Burlington Major Carlisle, Senior, was always *Mr.* Carlisle; junior was usually just Bud. Nowadays some club owners are still addressed by their keepers as Mr., but most—like the rest of California society—have a first-name relationship.)

New Threats

It was too good to last. Throughout California in the 1950s and 1960s, population pressures continued to grow and with them the danger to all the state's wildlife. The hunting community, buoyed by some great seasons, went into battle needing all the clout it could muster to oppose them.

The Spinks concept of a wide ditch from the Calhoun Cut to Montezuma Slough (Chapter 1) was revived in the 1950s as part of an even greater threat to the marsh: the proposed construction of a Shasta Pulp and Paper plant east of Fairfield, whose effluent would be carried away to the Montezuma Slough. This time it was simply voted down, but within a few years the Bureau of Reclamation revived Spinks' dream yet again, in the hopes of freshening Grizzly Island's water. The cost, however ($7 million versus $4 million estimated by its optimistic promoters), again proved to be the deciding factor, and the idea was abandoned, probably for good. Among other uncertainties in the project was the question of what the diversion of so much fresh water from the Sacramento River might do to the position of the "hydraulic barrier," the brackish separation between the salt and fresh waters that was so vital to the marsh's ecosystem. (Coon, *West Wind,* Apr and Jul 1994)

The biggest problem of the 1950s, however, was the 1958 Burns/Porter Act—the California Water Plan—that allocated water all over the state and each year reserved 4.25 million acre-feet behind the Oroville Dam for diversion through an aqueduct to the southern half of the state. Far down stream, this meant that the saltwater intrusion into the marsh was bound to get worse as the flooding tides from the sea were no longer opposed by the full force of the valley's fresh waters.

Suisun Resource Conservation District

There was no denying the menace that the Burns/Porter Act was posing, and on 4 June 1963 the majority of landowners in the Suisun area, mostly duck clubs, organized under the leadership of James Bancroft to form the Suisun Soil Conservation District, in 1964 renamed the Suisun Resource Conservation District (SRCD). The first board of directors included Bancroft, Ron Dysart, Glen Lawler, Stanley Anderson, and Dr. William Coon. Other founders were Fred Cutter, Herrick Low, Brooks Walker, Bill Towne, Dean Witter, and Gene Whittaker. They mobilized to protect and preserve the marsh and 27,700 acres of its flanking grasslands, in all an 89,000-acre tract that reached from Route 680 in the west to Shiloh Road in the east, and from Route 12 in the north to Suisun Bay in the south. Today the acreage has expanded to 116,000.

In May 1964, SRCD signed a memorandum of understanding with the US Department of Agriculture's Soil Conservation Service to delineate their respective responsibilities. Although set up in traditional soil conservation language for the "objectives of helping to bring about the conservation, development, and wise use of land, water and related resources," the real purpose was to preserve the marsh and its highly productive brackish-water environment for wildlife. This was finally spelled out in a 1980 SRCD resolution referring to its mission, in conjunction with the San Francisco Bay Conservation and Development Commission (BCDC), of preparing and enforcing "a management program designed to preserve, protect, and enhance the plant and wildlife communities within the Primary Management Area of the Suisun Marsh, including ... enforceable standards for diking, flooding, draining, filling, and dredging of sloughs, managed wetlands, and marshes."

Undecided at first, however (and still sometimes a source of acute controversy) were the best tactics and strategies for achieving SRCD's goals. On one side were those who wanted to go into full combat, fighting to the last political bullet to repeal the California Water Plan. On the other side were those who saw in such a hard-nosed approach the seeds of disaster. Given the preponderance of votes in the southern half of the state, said the second group, the only salvation was to negotiate the best possible deal, maneuvering within the fairly narrow bounds of political realism. Bancroft pushed for the second strategy, and his view prevailed. His thinking is reflected in this extract from a 5 July 1966 letter to Carl Wente:

> Dr. Ward made it clear that he thought we were making a monumental blunder in our approach. He told us we should marshal all our resources and work to repeal the California Water Plan. He had little interest in lesser objectives.
>
> Our own attitude was that we really had no hope, as a practical matter, of repealing the California Water Plan. Rather, we felt that we should marshal our resources to live with the Water Plan, trying to mold critical details of its execution so that we can live with it. Our consuming purpose was to have the Suisun Marsh well provided for in the California Water Plan and in whatever subsidiary plans might affect it. Also, we felt that because the U.S. Bureau of Reclamation was already here and in business every day, we should immediately try to establish a sound working relationship with it....
>
> My hat is off to Dr. Ward as a tireless crusader and solid citizen. He honestly felt that we were making a mistake in our approach. History may prove he was right. Many good men have lost a lot of sleep and gotten nowhere trying to do things by peaceful persuasion instead of by bulldozer. However, I just don't think that we or Dr. Ward had a bulldozer, and I'm not much for bluffing. (Dr. Coon files. Reprinted with permission of James Bancroft.)

In 1966 SRCD commissioned a study of the marsh that produced two "Final Reports" (Phase 1 and Phase 2) tracing the marsh's history and describing its biosphere. On the basis of these excellent reports, the Suisun Marsh Preservation Act of 1974, requiring development of a protection plan for the marsh, slowly took form and was eventually pushed through the California legislature.[5]

More Assaults

SRCD was formed none too soon, as more and more covetous eyes were being focused on the marsh's development potential. Most of these were corporate interests representing industry and agriculture, but the US Navy was also a player. There was even an absentee family group that had inherited some 1,500 acres and had visions of transforming it into some commercially viable real estate. The 320 acres that lay behind levees—most of the balance lay under water even at low tide—was being leased for a nominal fee by the Tip End Duck Club.

In June 1964, just a month after SRCD was formed, the heirs commissioned Philip H. Arend's Wildlife Associates, Inc. to do a land-use study of the property. Arend, whose other writings reflect an understanding of hunters' devotion to the wetlands, must have had some trouble keeping a straight face as he responded to their questions about this most remote corner of the island. No, it probably would not do for industrial development, which would cost $100,000 an acre at a minimum. No, the salt content of the soil would make agricultural development for farming also prohibitive. A marina? Well, possibly—at some far-distant future date. Not even improvement into a luxury duck club made much economic sense. All in all, Arend concluded, the minimal rent being paid by the present lessees

---

[5]This act owes its passage to many people but to none more than CWA's Dan Chapin, who lobbied tirelessly for it. When Governor Edmund Brown was in seclusion studying budget options in 1974, Dan got a call from Brown's harassed secretary. "Listen, Dan, no matter what I did to protect him today, seven calls got through, *all* of them about your goddamn marsh bill!! Yes, he'll sign it, but tell all those guys to lay off, will you?" (Chapin Interview, 1994)

for their primitive duck club was probably the best deal the family could hope for. (Arend, Land-use, 1964)

If dreams of developing Tip End could be laughed off (later, the lessees heard that the owners were even considering a shopping mall there), such was not the case with other, more accessible parts of the marsh. Over the next ten years, National Steel hoped to construct a plant on the east bank of Montezuma Slough near Collinsville, and planners of a nuclear power installation wanted the same property. A thermal power station was also planned. Dow Chemical had in mind a plant along the Sacramento River just upstream. The San Luis Drain proposed dumping agricultural waste waters into the delta, with the potential of afflicting the marsh with selenium poisoning, as happened in the Kesterson Refuge. For its part, the US Navy wanted to use the marsh as a training ground for riverine operations in Viet Nam. But the worst threat came from Envirosol, a disaster that nearly happened. (Coon, *West Wind,* Oct 1994, Jan 1995, Apr 1996)

Envirosol

Before the Suisun Marsh Preservation Act passed (and before SRCD had the teeth it does today), an unexpected menace raised its head. In 1973, Envirosol, a Seattle-based enterprise whose "green" name belied its true function of garbage disposal, chose the Potrero Hills as site for a mega-dump that would service all communities from San Francisco to Sacramento. Giant levee-threatening barges with several million people's refuse would ply the Montezuma Slough, disgorging their ripe loads at a pier where trucks would pick them up for interment in the hills.

Promoted by a smooth PR campaign, Envirosol at first enjoyed considerable local support, including that of the Solano County Planning Commission. Aroused members of the duck hunting community, led by Dan Chapin (CWA), Bill Coon (Vallejo dentist and duck club president), and Bill Frost (Grizzly Island rancher) appeared to be engaged in a losing fight to stop the project.

After a hearing held in Fairfield, however, help came from an unexpected quarter. One Ed Bruce, describing himelf as a Southern Californian in a "water-oriented industry" and president of the California Wildlife Federation, heard their arguments and offered to help crystallize opposition to Envirosol. Working tirelessly behind the scenes, he showed Dan and the two Bills how to create an effective political opposition, the Suisun Defense League, that mobilized Solano County public opinion and eventually produced a 12,000-signature petition that killed the project.

Unlike other Defense League leaders, however, he had a passion for anonymity, at one point even politely threatening to withdraw his volunteer services if his name were revealed. His compulsive modesty soon became understandable: he was president not only of the California Wildlife Federation but also of Sunset Scavengers, a San Francisco-based "water-oriented industry" that dumped garbage at sea and, unlike Envirosol, did not try to mask its true purpose in life. Neither, in private, did Ed Bruce. As he later commented to Dan Chapin, "We are not in the business of encouraging competitors." (Chapin, Coon interviews, 1994)

More Marsh Protection

The death of the Envirosol project breathed new life into the marsh protectionist movement. After the cornerstone Suisun Marsh Act of 1974, there followed a flurry of other plans and acts. The Fish and Wildlife Element to the Suisun Marsh Protection Plan was issued in 1975, and in December 1976 the comprehensive, 50-page Suisun Marsh Protection Plan was presented to the legislature. Next came the Suisun Marsh Preservation Act of 1977 and its incorporation into the Public Resources Code, and finally Water Right Decision 1485, issued in 1978 by the State Water Resources Control Board. The last-named report, which was modified in 1987, set maximum permissible salinity standards to guarantee the proper brackish water from October to May for Suisun's historic environment. (During the dry season, tidal salt water advances upstream beyond the marsh's boundaries, and freshness cannot be guaranteed.)

It was a good thing that the first defenders were there to take up the cudgels on behalf of the marsh. In the early 1970s good shooting and generous limits had again made hunting popular, leading to a small boom in club sales but also bringing to the marsh newcomers unaware of all the political complexities swirling around their properties. In 1972 Bud Carlisle sold the Teal Club to a bunch of youngsters fresh out of college with little or no experience in the game. Over on Grizzly, a new club called California Farms took over the old Puccinelli club, another case of energetic young men acting with more enthusiasm than expertise.

Elsewhere in California, real estate booms in wetlands and other undeveloped property had been sure signs of impending habitat destruction, and the same thing might have happened here. That it didn't is probably thanks mostly to the SRCD watchdogs. It is even possible that some new owners themselves figured that the clubs could turn into goldmines if the marsh were ever developed. (Few if any seem to feel that way now. In fact, it almost looks as if the marsh has developed the owners rather than the other way around. Those who were new in the 1970s are today's old-timers, now among the marsh's staunchest defenders.)

The 1980s and Beyond

Early in the 1980s came another plan for exploitation, this time involving windmill farms for power generating on the Potrero Hills. Again it was opposition from the duck clubs, which focused on the potentially murderous impact of the project on migrating raptors and other bird life, that stopped it. The development was relocated off to the east, where its impact on bird life may be less threatening.

In 1984, a multi-agency, comprehensive Plan of Protection for the Suisun Marsh Including Environmental Impact Report established step-by-step measures to be taken by each landowner and provided for ongoing monitoring. The plans for state-managed activities were most recently laid out in 1993 in the ponderously titled Screening Alternative Actions and Describing Remaining Actions for the Proposed Western Suisun Marsh Salinity Control Project.

There is no end in sight to these infinitely complex programs, which are also subject to change or abrogation as new species become endangered, unwanted environmental side-effects become apparent, or new private or public interest groups demand a say in marsh management.

Today SRCD is (or should be) the focal point for any activity involving preservation or enhancement of the marsh as wildlife habitat. Its present executive director, Lee Lehman, has been a member since 1965, a board member since 1980, its president for eight years beginning in 1985, and since then its fulltime manager. Most of the original founders have now died, but continuity is maintained through a rotating board of directors who own property within the district. SRCD holds monthly public meetings, often quite lively, where those with an interest in the marsh can hear the viewpoints not only of the duck hunting community, but myriad other associations.

Among the old and new groups who either have or want a say in what goes on in the marsh are the Army Corps of Engineers (management practices throughout the marsh), BCDC (plans), California Fish and Game Department—and the U. S. Fish and Wildlife Service—(hunting regulation), Solano County Mosquito Abatement Control Board (flooding and draining times), Bay Area Air Quality Control Board and Solano County Fire District (times and acreage of marsh burning), Department of Water Resources (water quality), California Waterfowl Association (habitat), Solano County Planning Commission (various construction), Citizens for Open Space, Suisun Marsh Action Committee, and, most recently, the San Francisco Bay Area Wetlands Ecosystem Goals Project. All but one of these groups has concerns other than the Suisun Marsh, but only SRCD has the requisite experience, institutional memory, and focus on the marsh to be an effective coordinator.

In fact, with this many organizations involved and such an overlapping of jurisdictions and responsibilities—some of them unilaterally assumed—it is somewhat surprising that anything ever gets done at all. Fortunately, that same Sargasso Sea of bureaucracy tangles those who would like to degrade the marsh for their own profit. Through it all, SRCD so far has somehow managed to navigate successfully, but the price of continued marsh security is eternal vigilance by everyone who holds these wetlands dear.

CHAPTER 7

THE ODD ECONOMICS AND SOCIOLOGY OF DUCK CLUBS

"Season closed March 15, 1882, with no more hunts. Expenses for the season—$93.60, average nearly seventy-eight cents ($.78) a bird, but Oh! The fun and excitement." M. Hall McAllister (age 21), hunting scrapbook entry.

"Great reluctance was encountered in attempts to elicit economic data....Many individuals either refused outright or procrastinated when asked how much it cost them to hunt ducks." (SSCD.1, p. C-30)

Costs and Finances

Some hunters, like the young McAllister, are bachelors. Others either have remarkable talents for concealing their finances or very understanding spouses. When it comes to duck clubs, it is a sport for the well-off or the very dedicated. Even those who hunt individually on public land spend more than they like to admit.

In the early days the price paid for renting hunting rights was usually fixed much as it is today, namely by the lessor's keen look at the prospective lessees and a wild guess at what they were willing to pay. There were big inconsistencies, and money was not always the main factor.

We don't know how much the Chamberlains charged the market hunters Payne and Beckwith in the 1870s, but it was surely no more than what Payne demanded in turn of the eight charter members of the Hardland Club in 1880: $75 for a six-month season on 210.8 acres, or less than $.04 an acre for a shooting year.[1] In 1881 the price had risen to $.06 the acre, but by the next year it had trebled for nonresidents while remaining fairly constant for locals and others with the right connections. Locals renting from locals continued to enjoy low rates all the way into the late 1890s. (CR)

The only available information on how the clubs valued memberships comes from a rare copy of the Cordelia Club's 1888 by-laws, which obliges the club to buy any resigning member's share at the fixed price of $200. It also gives the president the right to assign any price he wishes when selling a membership, which would imply that the club believed it could do better than $200 on each, or $2,000 for the club.

The main absentee landlord (the Chamberlains) at first seemed unaware of the rising value of their marshland, but after they saw their individual lessees (McPike and Richards) subleasing at a handsome profit, they caught on. For some unknown reason, however, they continued to give the Teal Club preferential treatment

---

[1] Measuring what this means in current prices is almost impossible, due to the vastly different conditions. A common laborer in those days earned $6.00 for a six-day work week that went dawn to dusk, or a bit over $300 a year. If we credit today's unskilled worker with an annual wage 50 times greater ($15,000 -- and that's probably high) and apply that to the first duck club rental, we get 50 x $.04 = $2.00 an acre, a remarkable bargain by any of today's standards.

for many years. Until 1893, the Teal enjoyed a flat $.07-an-acre rate that soon was only a quarter of what others were paying. Perhaps the political weight some Teal Club members threw around in Sacramento as well as Suisun had something to do with the matter. (State Senator G. Frank Smith had been a Teal founder, and Governor James Budd occasionally shot as the guest of Ed Goodall, also a Teal founder and a prominent Suisun figure.)

The Chamberlain estate managers generally did not gouge their tenants, but old Jabez Thickbroom had no such reservations when it came to city slickers. He rented out the 590-acre Tule Belle marsh for the outrageous sum of $.34 an acre in 1885 and had the gall to include clauses reserving all grazing rights to himself and making his lessees specifically responsible for any livestock wounded by them or their guests. But next year Jabez defaulted on his mortgage and lost the whole property, which his tenants then bought, becoming the first club on the Suisun Marsh to own property.

If marsh rentals were capricious in the 1880s, they were wild in the 1890s, ranging from $.06 to $1.15 an acre. Granted, the high figure was courtesy of Hall McAllister, whose enthusiasm for hunting in 1890 obviously overcame his financial common sense, because the next highest rent until the close of the decade was less than half what he paid. On the low end, politics probably outweighed economics. The infamous Mallard Club, whose members poached at will, had the cheapest rental, almost surely because their Suisun landlords were counting on them to make life miserable for the Cordelia Club. The Suisun Gun Club got away with a moderate rental of $.20 an acre because their membership was mostly local, as were their landlords. Did the Teal in 1893 lose its lease on purely economic grounds or did it offend the Chamberlains in some way? We don't know, but the Cordelia had to pay four times the Teal rental when it took over. (CR)

The cheapest rent was the $25 that Pete Smith paid Fred Goosen for one year's hunting on the latter's marsh in 1897. Smith eventually bought the marsh, now the North End Club of more than 400 acres, where the Smith family still lives and runs its heavy equipment business.[2] (Ibid)

As recreational hunting became established, tailored caveats began appearing in the leases. Some leases were specifically for hunting and/or fishing, most reserved the landlord's right of pasturage, and in one case a lessee was enjoined against extending a 500-foot ditch he had begun (probably for access to a pond) because it interfered with cattle-grazing. After the 1890s' trespassing disputes, tenants began demanding and getting written assurances from landlords that the latter would fend off unauthorized visitors. (But the tenants would be responsible for paying any prosecution costs.) (Ibid)

The Sell-Off Starts

Once the Dinkelspiel syndicate had broken the Chamberlain monopoly and opened up the marshes for subdivision and duck club sales in 1905, a land rush began:

> The amount of money invested by local sportsmen in duck clubs and preserves would about cover the sum total of a national bank's yearly operations, and the annual cost of maintaining them would run several small cities....
>
> Marsh land that ten years ago was regarded as not worth $5 an acre cannot now be purchased for $50 an acre.... Within two years, half of the [Chamberlain] tract was sold for more than the purchase price. The last open piece was sold a year ago, 251 acres at about $35 per acre to the Seymour Gun Club. The total cost of the land and improvements will reach nearly $20,000. Another sale recently was that of an 1,100-acre tract [*Author's note:* The Tule Belle] for

---

[2] The Goosens reportedly were a contentious lot. William Goosen regularly sued the Field and Tule Club for burning tules and destroying pasturage, even though the club successfully defended itself each time. "Goosen likes hitting his head against a stone wall," commented the *Breeder and Sportsman*. (*B & S,* 13 Oct 1906)

over $40,000.... (De Witt, p. 444)

The above article was inaccurate in many respects. The Seymour Club was not the last open piece by far, and the contentions elsewhere in the article that the Chamberlain tract was only 2,000 acres and sold for $120,000 were both wrong (its more-than-5,000 acres went for $110,000), to say nothing of the author's solemn promise that spoonies arrive on the marsh late in the fall in "succulent condition." (Sometimes, Mr. De Witt, sometimes.) If anything, real estate prices were going up even faster than the reporter stated. In 1912, perhaps inspired by a published story that marshland unsellable at $2 to $3 in the 1890s had recently been bought for $37 an acre by a man who was refusing $100 an acre for it, Arnold and Denman paid nearly $75 an acre for their marsh. (*B & S,* 10 Feb 1912; CR)

When that first real estate balloon on the marsh popped is not clear from the records, but given the Arnold genius for business, it was probably about five minutes after the last transaction above. Certainly hunting fell off about then, and land values must have fallen too. Later, the Great Depression of the 1930s sent land values plunging to all-time lows. In 1931, Andy Armstrong sold his 1,711.3 acres to the State of California (in part for its Joice Island Refuge) for $65,885.05, or $38.48 an acre, about half what Arnold and Denman paid for their marshes just across the Suisun Slough in 1912. Things really hit rock bottom in 1938, when hunters who for ten years had rented the Hastings marshes north of Montezuma Slough (the Greenhead Club), were able to buy it for $10 an acre.[3] (Steel, p. 10)

But after World War II, and particularly after the booming duck

---

[3] The Greenhead owners were twice blessed, paying the least for their property to begin with and then discovering that big gas pockets lay beneath them. From 1944 the members paid no assessments and in fact turned a profit on the club while Standard Oil paid for building and maintaining state-of-the-art levees.

flights of the 1950s and early 1960s, land values climbed steadily and sometimes spectacularly, with only occasional slips. One of the early good bargains (though it may have seemed exorbitant at the time) was the 1951 purchase by Fish and Game of 8,600 acres of Grizzly Island for $650,000, or $75.58 per acre. This is the Grizzly Island Game Management Area, which now provides a home for migratory waterfowl as well as pheasants and tule elk, and which is the site of a highly successful duck nesting program. Its public hunting areas are among the most convenient to the San Francisco-East Bay urban communities, and it provides other outdoor recreation, from fishing to bird viewing, all year round. (Boatwright, p. 16)

By the mid-1960s the price of marshland had risen to $250-$300 an acre for clubs of 300-400-acres In 1972 the Teal Club sold for $462.58 an acre, and in 1980 an appraiser suggested a market value of $1,200 an acre. By the late 1980s, the average selling price was $1,000 an acre. Given that this was also a time of falling limits and shortening seasons, club values held up well. (CR) If present trends in larger duck populations, bigger limits,and longer seasons hold up, land values could rise in the '90s.

Buying a club outright is usually more expensive per acre than buying a proprietary share in the same property. A ten-member club valued at $500,000, for example, will almost always value its individual memberships at less than $50,000. If a hunter finds a compatible group already in place and seeking new members, he can do much better than if he puts together his own syndicate to buy a whole club. But he must conform to an existing club etiquette and character, which is not always easy. Trial marriages of a season or more between clubs and prospective members are a common way of testing compatibility before signing on permanently.

Those who spend a great deal to buy all or part of a club usually do so with the consoling thought that private wetland habitat is limited, finite—and shrinking. No more is being created, and what there is should command steadily higher prices in the long run.

Early Taxes

In the early days, even before the clubs owned their land, some of them were taxed for personal property. The Teal Club's possessions in 1902, its first tax-paying year, totaled $400 in value, of which poultry and cattle accounted for $25 each; boats, furniture, and "FA" (Farm Apparatus?) were $50 each, and "improvements" (presumably the clubhouse/ark) were $200. Its 1902 tax bill, $6.60, was raised to a crippling $9.62 in 1903. Of course the intangible benefits of membership—and the price of joining—must have been considerably higher if the Cordelia Club's 1888 by-laws are any yardstick. (See p. 116.)

Three years later Herman Oelrichs declared $3,000 in improvements on the Tule Belle property, and in 1908 the assessed value of improvements on a dozen new and old clubs ranged from $0 to $3,000, with Achille Roos and Joe Harvey each matching Oelrichs' high figure. (One gets the impression there was an element of keeping up with the Joneses here—all three clubhouses were considered "palaces" at the time.) By then the rate of taxation had also risen: the Family Club, with $1,000 of improvements, paid $46.75, for example.

During this era, coincidentally or not, all of the clubs owned or leased by local residents were free of taxes, whereas all but one of those owned by nonresidents had to pay. (The one exception, the Volante Club, had been owned by nonresidents since 1907, but the founders of its predecessor, the Belvedere, had come from Suisun, and some of them were still Volante members.) Why the discrepancy? One can only guess that the assessor—operating without prejudice, of course—drew a distinction between those who owned only a valueless duck shack for changing clothes in the field (honest locals) and those who had a disguised mansion (them rich carpetbaggers from the city). Whatever the case, by 1911 the Volante Club was also being taxed.

As the 20th century unfolded, property taxes rose along with land values and at times took huge jumps. A 200-acre club bought in 1948 for $32,000 had $48 in taxes the first year, but in 1965 the owners had to pay $450. "Exorbitant taxation has been a contributing factor in the destruction of the wetlands of San Francisco Bay. Inflated land values and excessive taxation literally drove the Alvarado duck clubs of the South Bay to extinction." (Arendt, "Present Data on the Suisun Duck Clubs," SSCD.1, p. C-31)

Nowadays taxes are usually not a serious challenge to owners. Local authorities understand that the hunters bring money and employment to the community, and discouraging them through higher taxes would be self-defeating. Clubhouses are no longer the country palaces of Roos and Oelrichs, and though some still possess a fading grandeur and a handful show modern luxury, most are modest and utilitarian, like their owners. As for the great outdoors around them, it remains blessedly unimproved except for the clubs' access roads, levees, and duck blinds.

Peculiarities of Early Clubs

As we have seen, the first clubs were composed mostly of socially prominent executives in various fields, men who knew each other well and often held memberships in such other clubs as San Francisco's Bohemian or Pacific Union. In the next generation duck clubs tended to be made up mostly of individuals in the same white-collar professions—doctors or lawyers, for example. Still later, membership spanned a broad social spectrum from blue-collar to millionaire, not only from club to club but occasionally within the same club.

By today's standards, the early clubs had some unusual aspects. Most of them, irrespective of the acreage or number of ponds, had exactly ten founding members, a figure that held true for the Hardland, Cordelia, Tule Belle, Joice Island, and Seymour, for example. No known law or custom mandated ten, but it may have come from an instinctive understanding that factionalism could become a problem if membership went higher.

Factionalism is always at least a potential problem, even in clubs with only a few members. When the number climbs above ten or twelve it is almost inevitable. Possibly in recognition of that danger, the Cordelia Club in 1888 set its maximum membership at twelve, and it later actually shrank in size, as did all the other early ten-member clubs. In the early 1900s, none of them had more than five members.

A survey of 85 Suisun Marsh clubs in 1953-1956 showed an average membership of seven, though the state-wide average at that time was ten. (SSCD.1, 1966, p. C-26) There has been no known census since that time.

Another oddity in the early days was the way members switched allegiances. McAllister and Babcock, for example, helped found the Cordelia Club in 1880, moved over to the Ibis in 1883 and 1887, respectively, and then later moved back into the Cordelia, decoying with them Charles Eells and James Otis. Louis Titus owned the Cordelia marsh in 1906 but was a shooting member of the Teal Club. Bill Richards and Bill Bradford flew all over the place. Now, such defections, redefections, and split allegiances are almost unheard of. These quirks probably just reflect the growing pains and uncertainties of a newly popular sport.

Certain other aspects of duck club sociology, however, have remained constant down through the years. From the outset, the Teal Club could never be confused with a monastic order of sportsmen. The hunting parties thrown there, hosted in the beginning by Commander Floyd during the 1880s, were unique. Every fall at the end of October, he and some of the other members brought their wives up for a week-long bash. "A special French chef with extra servants went along," recalled a nostalgic McAllister in 1930, when he was nearly 70, "and they also took with them the famous old-time `Macaroni Band,' an Italian orchestra that used to play on the Sausalito ferry. The men would take the morning shoot, the ladies of the party would go with the keepers for the afternoon hunt, and in the evening they had music and cards.... There were no limits on ducks, champagne, and cards."[4] (McAllister, 1930)

But the allegation in the 8 September 1893 *Examiner* that each Teal member and guest had three servants in the blind—one to load and clean the guns, one to hold a sunshade, and one to "fill the air with perfume" from an atomizer—was a bit overstated.

Club Size and Club Character

The 153 clubs on the Suisun Marsh vary enormously in size, from the one-man, 5.25-acre Machado Duck Club, to giants like the 1,311-acre Joice Island Gun Club. According to present zoning laws, the bigger clubs may not be broken up into anything smaller than 250-acre units. This has prevented subdividing and guaranteed a fairly stable set of property lines. There are about two dozen clubs with more than the 500 acres required for a split, but probably no more than a handful could be broken up without extensive levee work to ensure independent water control over each part. And anyway, only the direst threat of bankruptcy or the bitterest of internal conflicts could convince most clubs to take such a step. Most have existed so long that they have become institutions, acquiring personalities as unique and radically different from each other as those of the collection of men who make them up. The acreage of a club is no more material in determining its character than the waistbands of its members indicate theirs.

Over the years a club's character may change slowly. The original founders may all have come from one profession, but as they

---

[4] McAllister's opinion should be respected, at least about the cards. In 1902 he staked his house in South Park in San Francisco on a second-best poker hand, losing to Captain Marty Lyon, who gave the house to his daughter as a wedding present.

grow old and retire, those they nominate as replacements are usually chosen more for their hunting enthusiasm (and occasionally for their mechanical skills and equipment) than for their professional fit. This means that duck clubs (unlike guilds or country clubs) throw together people from widely separated walks of life. Whether the experience proves devastating or stimulating for club unity depends on the flexibility of its members.

Only if a club is sold outright or undergoes a swift wholesale change of membership, however, is one likely to see a sharp change in group character. Even then, there may be certain carry-overs, such as the keeper or even just the clubhouse, that imprint forcefully enough on the new owners to preserve some of the old. It is hard to imagine the Montezuma Club, for example, abandoning its dedication to casual rollicking good times and relaxed rules any more than one could expect uptight, formal clubs to escape theirs. Clubs must, of course, undergo periodic rejuvenation or they will die. But rejuvenation does not necessarily involve revolution, no matter how much some of the more rigid ones may need a shaking.

Ownerships and Stewardship

There are six classes of marsh ownership, ranging from total to zero, and the first four are directly related to how we care for the land.

First comes the single owner, who is either extremely wealthy, or owns a relatively small tract, or has seasonal shooters or leased blinds to defray his costs. The single owner has all the power, but he also has all the responsibility, and he almost always sees to it that his marsh is properly looked after. His should be—and usually is—a permanent, individual commitment to the property, but he must take on responsibility for the actions of any tenants as well.

The second, more common type of ownership is the joint proprietary partnership. Here we have shared responsibility and commitment, a more chancy proposition. The members are locked in to some degree—it is not always easy to sell a proprietary membership if one wants out—so they must ultimately pay the consequences of neglect or deferred maintenance. But unless there is enthusiastic cooperation among them, the diffusion of responsibility can result in a "let-George-do-it" attitude that leads to long-range problems. Club spirit—or the lack of it—usually spells such a club's salvation or ruin. Again, seasonal shooters may be let in to help carry the financial burden.

The third kind of ownership is the joint leasehold. Each joint lessee has a shorter-term perspective than a true owner. He is able to bail out on little or no notice at any time, but he also runs the risk of being expelled by an absentee landlord who wants his land back. The ties among members within such a club tend to be looser, the commitment to the property less intense, individual attitudes more selfish, and the interest in good relations with neighboring clubs lower. It is a cheaper way to go but less secure.

In fourth place comes the hunter who is either a seasonal member (with some or all of the rights and obligations of a part-owner except on policy matters), or who simply leases a specific blind. Especially where only a blind is involved, personal responsibility must rest on his conscience, not always the most reliable foundation. The blinds available for any newcomer are almost by definition the least desirable, which can lead to hard feelings between lessor and lessee that sometimes degrade the whole club. Unless the owner(s) are selective in their choice of lessees and establish clear codes of behavior, hunting may be spoiled for everyone around: sky-scraping shots, shrieked abuse among lessees, disputes over downed birds, and littering of the marsh, to name but a few. And whether well-behaved or not, lessees have the least incentive to preserve the environment.

Fifth comes the hunter on the state-owned refuges. He pays his daily or seasonal fee to the Fish and Wildlife managers for the privilege of hunting in competition with others on the same marsh.

Most of the hunting is free range, which means that the refuge shooter must depend on the good manners of his fellow hunters to a much greater extent than he would in a fixed blind. But if nothing else, he can have the confidence that his marsh is given the best possible habitat treatment. And despite occasional misbehavior, there is a certain communal spirit that builds up among the public-land hunters, especially those who hunt often enough in the same areas to get acquainted.

Sixth and last, there is the rugged individualist who owns no property at all, not even a day's lease, and who depends only on himself. This is the slough shooter, the sculler, and the freelance hunter on the few remaining public tidelands. As the least financially committed, such hunters might be expected to be the least responsible, but that is not the case. The few I know personally are the most careful nature observers and the most expert wildfowlers of all—and they keep the best records.

And then there are the poachers—but that's for the next chapter.

But, to end on a philosophical note, the real ultimate owner of the Suisun Marsh will be the waters that one day—a hundred, a thousand, ten thousand years from now, they don't care—will breach the levees for the last time and reclaim their own.

Rules

A good way of assessing a club's personality is by examining its rules—and the members' observance of them. Generally speaking, a club with paragraph after paragraph of detailed regulations is a club that does not trust the good sense and good manners of its members. A club whose members regularly watchdog the others' observance of every letter of the code is likely to be a happy place only for people who crave structure above all else. By contrast, the club whose camaraderie lets it get by with few rules and lackadaisical enforcement of the few there are may be somewhat chaotic, but it suffers far fewer ulcers.

Every now and then, however, one encounters an unusual hybrid between the two extremes that defies definition. One club was so stuffy that it had a firm rule that members would only be served dinner if they were suitably attired, which was defined as wearing a tie to the table. In revolt, two of the members began complying to the letter. That's all they wore. (I never did hear how that one turned out; surely one side or the other must have prevailed.)

One of the key rules is how blinds are allocated. At a single-owner club, of course, all rules, including blind assignments, are at the whim of the owner, who can change or ignore them whenever he wants. Such was the case with my father, Big Stan, who personally assigned each shooter his blind at breakfast before each shoot. In theory the best blinds rotated among members, but exceptions were always being made for one reason or another. A bad shoot on a previous day for one member, for example, or the need of another to give an out-of-town visitor a memorable experience would lead to their getting the best choice. It was all done with easy but gentlemanly informality, and if anyone had objections he kept them to himself. At that time Will Richards' Green Lodge may well have operated in the same way, but nowadays different rules usually apply.

In some clubs there is no choice. Each member has his own permanent blind, meaning that the newest member has the poorest choice and must slowly work his way up the ladder as old members die or resign. It is an ideal system only for those who have reached the top. More commonly, there is a system for each member to choose his blind on any shooting day, with the priority of choice either a matter for gambling the night before or established as a rotational order before the season begins. Sometimes such rotational systems can be quite complex in order to prevent any one hunter from being advantaged or disadvantaged by long stretches of good or bad hunting weather. Mother Nature has never believed in even playing fields, however, and complicated formulae to correct her ways almost never work.

Other important rules include guest privileges (which range from none at all to a wide-open policy) and shooting days, which at first were limited to Wednesdays, Sundays, and holidays, and in modern times have included Saturdays as well. Our earliest club rules are again provided by the Cordelia Club's 1888 by-laws, which though few in number seem surprisingly restrictive for the time: no more than one male guest per member per day, a guest charge of $2 a head, the obligation for hosts to accompany their guests personally, no more than two invitations to the same guest per season, no shooting allowed on Saturday, and no shooting on club property for more than two consecutive days in any one week. (At that time, given the general custom, hunting on two consecutive days would only have been permitted where holidays fell on the right day. Under these rules, however, limited off-day shooting appears to have been possible.)

The next available rules are those of the Volante Club for the 1911-1912 season. They are far more elaborate than the Cordelia's and provide an early example of overkill. Spanning four tightly spaced pages (and entitled *extracts* of by-laws and rules at that), they cover everything from how to hold a decoy when throwing it out, to a ban on tipping the help. Like the Cordelia before it, the Volante charged for guests ($5 a day, or $7.50 for two consecutive days), and it had tighter guest limits: no more than five guests for each member during an entire season and no more than five visits by one non-member, even with different hosts. Various fines were mandated for noncompliance with the rules, including $1 for failing to fill out the register of birds taken and $5 for using another member's room or equipment without permission. Shooting days were Wednesdays, Sundays, and legal holidays from October 15 to March 1, and discharge of firearms at other times was strictly forbidden. To fire a rifle or pistol was grounds for suspension (the only listed transgression bearing such a penalty). The shooting day for ducks ended at 1:00 PM, except for the first and last days of the season, when it ran all day.

Some of the rules suggest low-class behavior ("Spitting on floor or in boats prohibited;" "(Please) leave (your) empty bottles unbroken in barrels"), while others hint of past catastrophes ("members are particularly urged not to leave matches loose in their clothing"). As a parting note, the last paragraph advised members and guests to familiarize themselves with the game laws of California, advice the drafters themselves may have ignored. The last shooting day, according to these rules, was March 1, 1912, yet the 1911-1912 season actually closed on February 15. In fact, the last time a season had ended on March 1 was in 1897, well before either the Volante or its predecessor club, the Belvedere, existed.

Of Women, Juniors, and Bonding

As the Cordelia's by-laws indicate, hunting was a purely male occupation in the old days. Most clubs probably did not even consider it necessary to spell that out. My mother, for example, never set foot on the old Arnold Ranch, with which Big Stan had been associated for nearly 40 years, even longer than he had known her, nor did she have any desire to. (I remember my vain attempt to get her up there after he died, in the hopes somehow of saving the place for us, but she would have none of it.) My diminutive aunt, Adaline Kent—an absolute dead-eye with rifle, pistol or shotgun—was probably the only woman ever to shoot that club. We boys, however, were welcome as soon as we could shoot, which for me came at age 12.

The Volante Club's rules, comprehensive though they were, did not mention women, but they did spell out rules on the young. No one under 14 was allowed, those from 14 to 16 could be taken out by their fathers without any assessment for guest privileges (but their birds had to be included in their father's limit), and those over 16 had to be treated as regular guests. All minors, however, were banished from the clubhouse living room promptly at 8:00 PM during the hunting season.

Over at the Wheeler Island Club, the adult-male-only rule held from its incorporation in 1924 to at least 1971. Then an otherwise very acceptable prospective member called Eastwood brought his 11-year-old son along for a shoot and so thoroughly outraged the older generation that a verbal donnybrook ensued. Feelings ran so high that Eastwood told them off in words of one syllable before departing for good. His parting shots on the value of hunting as a part of father-son bonding were effective enough, however, that the club soon changed its policy. It even eventually instituted a Ladies' Day, which has since become a regular feature of the season there.

The Honker Club had a rule that restricted guests to a member's *oldest* son, who was allowed *one* shoot per year. Some clubs might have considered even that unacceptably liberal.

Few if any clubs are completely coed, and Ms. Davisson (p. 85 footnote) is nowadays the only lady known both to own and shoot her property. Most clubs, however, have either designated ladies' days or rules that permit women at specific times, usually Sundays. This is a tradition that goes back at least to the 1920s, when one club had a ladies' weekend that involved a measure of practical joking. It seems there was an odd arrangement for bathing, whereby the bathtub in the ladies' facility could only get cold water through a pipe that led from an open tub on the gents' side. The younger male members delighted in putting small frogs in this receptacle, and the frogs would pop through the faucet on the ladies' side any time one of the bathers wanted cooler water. Audible through the walls there usually came a satisfactory (for the boys) reaction.

One of the more liberal clubs on woman guests is the Teal, whose only restriction is that Friday nights are exclusively for men, but women may hunt the club on any shooting day. There are, however, certain perils in this. We have cited instances where duck clubs have served as a cover for male philandering, but many years ago the wife of a club member was a bit too successful as a hunter, having set her sights on game other than ducks and having been caught by her spouse *in flagrante* with another member when she was supposed to be out in a blind. Pity. The club lost not only two members but a regular attractive guest as well.

Surprisingly, bonding across generations is not as common among hunters as one might think. Many in the field today did not inherit their passion for the sport from their fathers, nor have all of them succeeded in passing it along to their offspring. My three brothers and I certainly inherited ours—from both sides of the family— but only one of the next generation is really an addict. (That's young Pete Arnold, as avid a hunter as you'll find.) My own three sons *like* hunting, but more, I think, because it gives the old man such pleasure to have them along than for what they themselves get out of it.

This failure to carry over to a new generation is not peculiar to America. Only one or two of the British visitors we have entertained each year since 1980, chosen at random from the English hunting population, learned shooting from his father, though most with sons are trying to pass it along. Where a father and son do share a love for hunting, I don't think there is a stronger bond that exists. Too bad it's not more common.

Bonding among club members is variable. Some fast and enduring friendships are made, but often the only point of contact for a whole year between any two members is the club itself. This may occur only during the hunting season, but most now have one or more work days during the spring and summer where members share some sweat in maintenance and construction work. Work like that is a great leveler and bonder. Where members are allowed to pay a modest fine in lieu of putting in work time, however, wholesale cop-outs and a resulting loss of club spirit can occur.

Out-of-season contacts between members of the same club may be rare and individual members completely unlike each other, but the common devotion to hunting usually overcomes differences in background, profession, personality, education, income, and even politics. Most vital of all in determining how people get along, however, is how each individual blends with the club's character as a whole. Over time, clubs develop a composite personality that is as real and unique as that of any of its members.

Interclub Relations

Between adjacent clubs, *really* good relations are not too common. We recall the early Teal-Cordelia feud, and in later years the Arnold Club maintained only guarded connections with the Teal, right across the tracks. Recalls my brother Stan,

> Once or twice a year, there would come this unbelievably loud outburst of drunken revelry from across the tracks, a clear sign that the party had spilled out the front door and was seeking new sociability. Usually the gentlemen of the Arnold Club [Note: not averse to a noggin themselves, but rarely noisily] would sigh resignedly and put on the masks of genial hosts, pouring a drink or two for the invaders before sending them back across the tracks—but I recall that one time Big Stan or some other member greeted the outburst with a 'For God's sake turn out the lights!' It did no good. Indeed...one of the Carlisle gang fell off the bridge into the slough one night and had to be rescued.

(The remnants of the bridge still stand, a good 6-8 feet above the water. That tippler was lucky to survive.)

Despite—or perhaps because of—being close neighbors, the Arnold Club had some dark suspicions about what went on over at the Teal. Occasionally Mr. Carlisle would hold a party for ladies, who would arrive by train with great giggling. The Arnold Club's disapproval was increased by a rumor that the keeper's duties then included trapping limits of ducks, wringing their necks, and sending a load of No. 6's down a row of them as they hung from a clothesline. Alas for a good story, some 60 years later I found evidence at the Teal that the girls were secretaries of Mr. Carlisle's firm, and the whole "orgy" was chaperoned by Mrs. Carlisle. Ah well.

Carlisle's Wild Wild Women Gather for an Orgy at Teal
(How Big Stan could have suspected this group passeth understanding.)

Like villages in primitive societies, communal friendships, if they develop at all, are likely to leapfrog over the closest neighbors to those one step further removed, if for no other reason than mutual distrust of the one in the middle. For those adjacent to each other, interclub suspicions and name calling, deserved or not, cover a variety of opposite sins: social snobs/social boors; "unprofessional"/"too purist;" bad management/secret baiting; lousy shots/game hogs; skyscraper shooters/sour-grapes complainers. And of course the blame for flooding, which is always ascribed to the neighbor.

In old Teal Club files is a classic letter written by a Mr. Robert Covey, a neighbor now deceased, responding to a complaint about his responsibility for flooding at the Teal and Rousseau Ranch (now Tule Hilton) clubs. Noting that 90 percent of all the clubs in the marsh were under water at that time, the writer conceded that the improper management of which he stood accused might have been justified "providing I had the ability to command the Heavens and the Tides." Covey then went on to deplore the Teal Club's alleged lack of cooperation in facing joint problems and other assorted sins.[5]

Feuds and vendettas will probably endure as long as the clubs themselves, but they are a luxury that the hunting community can no longer afford and would do well to put aside. Beset by foes who range from animal-rightists to commercial developers to primeval restorationists to species-specific advocates (harvest mousers, delta smelters, Suisun pea-ers, etc) to rampant government bureaucracies to Mother Nature herself, hunters in general—and on the Suisun Marsh in particular—must unite if they are to survive. Clubs do have

---

[5] Under similar flood conditions, the old Tip End Club (not, admittedly, known for its own cooperation with its neighbors) was threatened with a lawsuit in 1983 for having flooded the whole south end of Grizzly Island due to poor management. It could have used Mr. Covey's rapier pen.

the option of banding together in associations to solve common problems, either by forming reclamation districts or creating joint-use facilities. These are options that offer significant benefits in exchange for yielding some club sovereignty, and they make very good sense for groups of clubs faced with problems too big for any one of them to handle alone. Joint action has become more popular in recent times, and it may be the wave of the future, but no one should underestimate the unwillingness of clubs to sacrifice their independence for the common good. Is this irrationally selfish? Yes, but it's not unusual. Look at the United Nations.

Life Cycle of a Club: The Poop Deck

Like people, clubs are born, develop personalities, mature, become jaded, and die. Sometimes they may have a rebirth under a completely new management, with or without a name change to signal a break with the past, and usually there are carryovers from the past and to the future that fuzz a life story. The Poop Deck Club was an exception to that rule, a club whose birth, life, and death were clearly defined.

Joan Frost, whose pictorial history of Grizzly Island was one of the inspirations for this book, was raised on a moored stern-wheeler ferryboat called the *T. C. Walker*. After her parents moved away, the boat was beached and abandoned on Bradtmoor (later, Bradmoor) Island, which belonged to the Lawler family. Two groups of local duck hunters used the hulk for shelter, shooting the neighboring ponds according to some informal agreement among themselves and with the Lawlers, but if they had names for their "clubs" they are lost to history. By 1944 only one of these groups remained, occupying the lower deck of the old ferry.

In 1944 three members of a deer hunting club dating back to the 1930s, Fred and Bob Cutter and Pres Snow, arranged with the Lawlers to move in and begin building a club from scratch. They started with a clean slate in the summer of 1944: no furnishings at all on their deck of the *T. C. Walker*, no serviceable duck blinds, no equipment. Their pond had not been shot for many years. About all they had was their new name—The Poop Deck—and the habit of keeping a more or less careful log.

The first season was a very meager one (so were the entries in the log), but the hunters did get a clubhouse of sorts going on the hulk's upper deck. In 1945 the club grew to 11 members, five of whom were Cutters, and the last remaining local hunters moved out. The members worked hard during that summer, and it paid off. Though elsewhere the shooting went downhill in 1945, the Poop Deck's record improved from a 1.8 to a 2.6 average.

THE POOP DECK CLUB IN ITS PRIME
(from a painting by Bruce Lattig-- courtesy of Ed Cutter)

More than that, the first club legends were recorded: A boatload of five hunters submarines, dunking all hands. — Ted Cutter climbs into a boat that he'd moored at high tide and goes aft to check the motor. The tide having gone out, the boat is only balanced on the bank until Ted's weight sends motor, boat, and owner into the drink.

The 1946 season notes were brief and to the point, e.g.,

"November 30: (members) give levee shooters hell and shoot up their decoys.

"December 1: Decoys at Lyon Pond stolen."

There were some brief allusions to hard drinking and its unhappy morning-after consequences, as well as the first mention of a guitar and singing dirty songs as part of the socializing. References to both quickly became standard entries in the log.

From 1946 to 1948, the average hunter/day kill climbed from 3.2 to 3.9 to 7.1, the last a level it would never approach again. This quick improvement gave the club a robust childhood, despite a poor opener in 1950, followed by a disastrous flood that brought the season to a premature end in early December. But the club survived with good cheer and optimism, as reflected in the following years' growth of birds shot and total hunter/days, which peaked in 1952 and 1953, respectively. Fishing for stripers and catfish became a profitable sideline, both during the season and after workdays, which began as early as April and continued through the off season.

Legal niceties were not always observed. The club noted that a neighbor to the north baited extensively and successfully with tomato pulp but only much later hinted that they were guilty of the same practice. There were two instances of hunters reducing a "swoose" to possession and several late-season long-tailed "rabbits" provided excellent feathers for fly tying. Limits were acceptable as general guidelines only.

Mainly, though, it is not the hunting but the growth of the club personality, one of fun-loving fellowship and a marvelous sense of humor, that shines through so clearly in the log. The club would hold from two to five work days a year in the off season, occasions when the members and their wives and children would all gather for dinner and a rousing party the night before and go to work the next morning.

They got a lot done, and they obviously had a good time doing it, despite biting comments and retorts entered in the log. To make the log acceptable reading for wives and children, some language was cleverly edited. ("Fred came in with leaking waders. Changed. Went back out. Fell in crack. Said, 'Gracious me.'") Other standard cussings-out were reduced to partial lines from old jokes (ARGO) or acronyms (POY).[6]

After the 1951 April work party, the log records that two of the members caught three catfish and two hangovers. That was when the *T.C. Walker* got a new roof and when the first recorded game of Pedro started a new Poop Deck tradition. Poker was already established.

Year by year, the physical premises improved. The survival items of beds, chairs, table, sink, camp stove, lanterns, and cold box were replaced and upgraded. The Standard Oil Esso Company drilled for gas, found none, but left a nice all-weather road and improved bridge behind them. Gas lights went into each cabin, then electricity. A new water tank, pump, and plumbing were installed.

The log itself improved, with better writing and more of it. All the usual hunter problems are there: the forgotten gun, the wrong shells, the morning wake-up alarm clock mistakenly still set at 1:30 for

---

[6] Anyone asking for definitions of these terms—and there were quite a few—immediately qualified for an entry in the log. There is no indication that their curiosity was ever satisfied, however. At the risk of breaking old Poop Deck security, let it be known that ARGO was the first part of what one might be told to GO do, whereas the second part of the POY acronym stood for On You.

the afternoon nap, occasional cases of obnoxious drunkenness (but never written with really hard feelings), dog troubles, dunking troubles, car-in-ditch troubles.

Then *real* trouble. On September 21, 1958, the *T. C. Walker* caught fire and was saved only because one of the members "was stupid enough to fight the damn thing." A bucket brigade did the trick, and by the time the fire department arrived, two hours later, there were only embers to put out. The fire chief almost poisoned relations by implying it was a false alarm, but the liberal application of alcohol to both the amateur and professional firefighters prevented a bad scene. (This is one of the longest and most exciting entries in the log.)

Exactly a week later, sandwiched between the fire and a routine entry on the forthcoming season's club members and limits, came a note that one of the club's original founders, Pres Snow, had died. Some other old members went up to the club for the sad task of seeing which of his duck-hunting possessions his widow might like to keep. "About all we could find were a couple of patchwork comforters his mother had made for him."

Entries over the succeeding years are still full of spirit and humor—the all-night card game, the problem of getting a wayward calf out of a duck blind barrel, detailed descriptions of poker and Pedro hands, heroic drinking bouts (including one where eight hunters consumed seven fifths of whiskey in one night)—but in 1961 a watershed was crossed when Bob Cutter took a membership in the Montezuma Club.

The next year a couple of other Cutters shot at the Montezuma, and thieves broke into the Poop Deck, stealing all 12-gauge shells, thermal underwear, liquor, cigarettes, and boots. It was a year of heavy rains and failing levees, and the club had to shut down before Christmas due to flooding, "one of the worst and wettest seasons in the history of the club."

Before the 1963 season opened, someone stole all the

canned goods from the emergency supplies, and over the 1964 New Year's holiday, a thief broke into the gun room and made off with a shotgun and some waders. The levee repairs of the previous summer turned out badly, and renewed heavy rains made for poor shooting

Despite these misfortunes, the old spirit still shone through. In 1964, the opening day dinner saw eight hunters tuck into 16 teal and 6 gadwall and wind up with only two birds as leftovers. But it is somewhat shocking to learn that one member put in his first appearance at the club in four years. In 1965 the limit dropped to five birds, and a run of bad weather resulted in the worst bird total (139) since the club began. It also finished with a dramatic fall by Ted Cutter off the stern of the *T.C. Walker*, a full two stories down to the marsh below. Incredibly, he survived unscathed, and at the opening of the 1966 season Rob Cutter solemnly presented him with wings in commemoration of the flight.

That 1966 opener was the most productive in ducks since the club began, and the same claim was made for the start of the 1967 season. But four members dropped out in favor of the Wheeler Island and Montezuma Clubs, and early in 1967 two of the Poop Deck's founding members, Fred Cutter and Walter Bergendorf ("Bergie"), died within a few weeks of each other. In 1967 and 1968, there were more cases of vandalism and theft from the club during non-shooting days or the off season, to the point that all the decoys were removed for safekeeping at the Cutter laboratories by open pickup—with a stop en route at a convenient carwash where the cargo was given a much-needed post-season sloshing.

Despite a promising beginning with a 25th anniversary party, the 1969 season was the worst on record, and the landlord, Glen Lawler, came in for unrestrained abuse for not living up to his marsh-managing commitments. ("POG! POG! POG!" read one entry.) The next season, 1970, opened bravely with a productive work weekend in September and a middling good opener, but Lawler was said to be trying to sell the whole island to the Nature Conservancy, and there was considerable gloom. The club still hung in for 1971, however, and that season became the best since 1960. The members believed that because Lawler had not grazed his cattle on the land during the summer, there was more feed for the ducks in the fall.

They also renewed for 1972, but in August Glen Lawler died, and the island passed to a relative, John Lawler, who seemed to have no intention of maintaining the property at even Glen's level—which the club considered inadequate. At a January 1973 meeting, the decision was taken not to renew the lease for the 1973 duck season. A final party, dismantling, and distribution by auction of all valuables on the *T. C. Walker* was held in May. "We didn't wash Friday night's dishes, just threw them overboard. Finally, we departed around 5 PM. There was no more booze left, so there was no reason to stay. It was odd, but we had a hell of a good time bidding the club farewell."

That should have been the log's last entry, but there are two more. Under the dateline July 8, 1973, it was noted that the *T.C. Walker* had burned totally on that day, probably thanks to two young vandals. The entry for that date closes with, "At this point one is tempted to write THE END at the bottom of the page, but I have a feeling we'll get together again to relive old memories, so perhaps the book will have future entries in it." And it did, but only one, dated August 9, 1973. "Bob Cutter died today." The last of the original founders had survived his beloved ark by a month and a day.

The remaining Poop Deck members dispersed to other clubs and only rarely reassemble. But to the extent that they have kept alive the old club's heritage of good humor, high living, and dirty songs, the Poop Deck still lives.

✧

Decline and Fall of the Poop Deck (1967-1973)

CHAPTER 8

LAWS, OUTLAWS, KEEPERS, AND WARDENS

Early Game Laws and Club Customs

California game laws go back to 1852, and fish and game commissions to 1870 and 1878, respectively, but it was not until 1895 that county boards of supervisors were authorized to pay wardens. (They were hired on two-year contracts at $50 to $100 a month and a maximum of $25 a month for expenses.) At that, by 1900 only six counties in the state had wardens, and Solano wasn't one of them. (Welch, 1930) By 1910, there were 100 wardens on the payroll for the whole state, but most of them were just deputies, and in 1916 the number had shrunk to 68. (*B & S,* 8 Jan 1910, 30 Dec 1916)

Some aspects of the old Fish and Game codes are quaint, such as shooting in a public cemetery, a misdemeanor that leads the list in all the early editions from 1872 onward. Others reflected a different level of concern for certain species than what is common now. The first duck seasons referred only to mallard and wood ducks. Later, in 1861 (possibly 1869), the law was broadened to include teal and spoonbill, and "any broadbilled duck." Rails had a closed season in the late 1880s, when ducks had a year-round open season.

Until well into this century, the open season was defined by the closed season. Certain kinds of birds could not be shot from a starting date in the spring to an ending date in the fall. Because the seasons were written into the legal code, they held constant until changed. From 1884 to 1887, ducks were not even listed among those protected in the closed season, which implied that there was no closed season at all. The history of the seasons is summarized on p. 147 below and laid out in more detail in Appendix C.

Most people think of our present concern for the environment as a recent development, a needed correction to the ruthless exploitation of nature that went on in the 19th century. It may come as a surprise that the first California laws prohibiting water pollution go back at least to 1872 and were written into the Fish and Game laws by 1914. "Every person who places...in any of the waters of this state...any lime, gas, tar, cocculous indicus (?), slag, sawdust, shavings, slabs, edgings, mill or factory refuse, or any substance deleterious to fish, is guilty of a misdemeanor, and is punishable by a fine of *not less* than two hundred and fifty dollars, or by imprisonment in the county jail...of *not less* than one hundred and twenty-five days, or by both.... all fines...shall be paid into the state treasury to the credit of the Fish and Game Commission fund. [Emphasis added.][1] (F & G Code, 1914-1915)

The first known arrest for hunting trespass near Suisun was

---

[1] Enforcement, of course, was another matter. Despite the fixing of minimum rather than maximum penalties (which implied a serious intent to apply punishment where warranted), there was apparently little effort to follow through. A 1905 article on oil refinery and paint factory pollution in San Francisco Bay gave horrifying details on the waterfowl suffering that followed, but there was no indication that the offending polluters were haled into court. A similar account followed in 1910, with special emphasis on the oil refineries. (*B & S*, 28 Jan 1905; De Witt, p. 443)

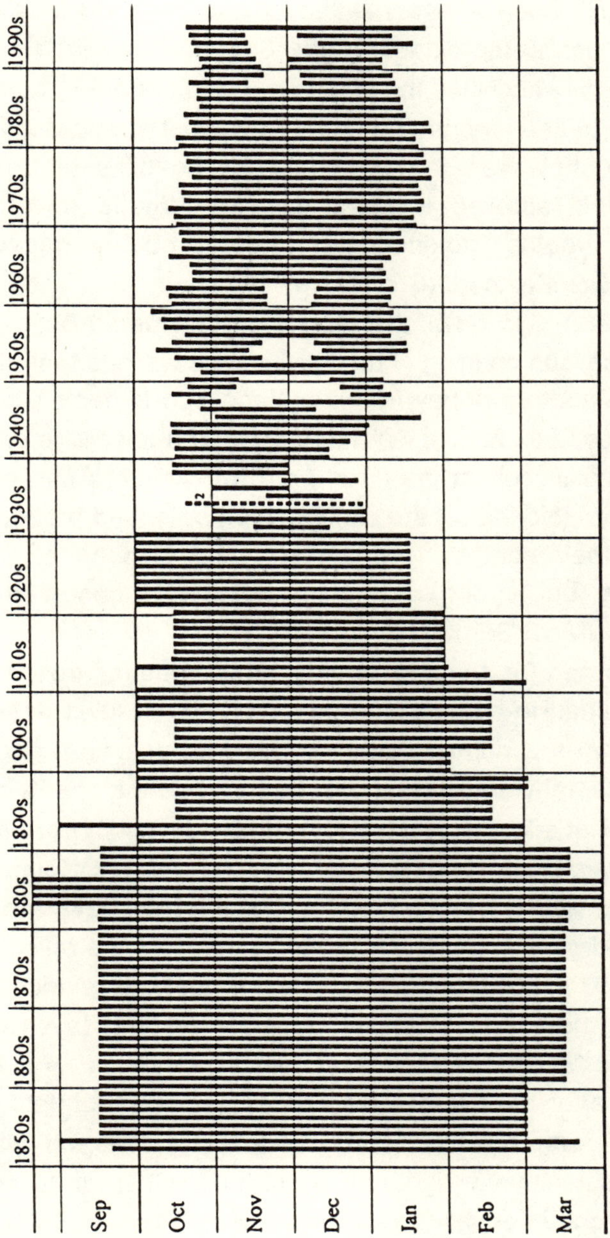

CHART 1 -- SUMMARY OF CALIFORNIA HUNTING SEASONS – 1852-1995

1. No officially closed season 1883-1886.
2. Sat/Sun hunting only in 1934.

logged in 1879, when an unnamed boy was caught on a local owner's property. Much to the disgust of the *Solano Republican's* editor of that time, the jury acquitted the offender. ("It is doubtful," complained the editor, "if ever a jury can be found to convict trespassers.") Only a year later, however, the same paper's headline of "Attempted Cowslaughter" showed a more cavalier attitude toward some trespassing youths' potshots at cattle, probably because the landowner was Mexican. (*SR,* 20 Nov 1879)

In 1880, two boys shot ducks on the marsh before the 15 September season opening. The culprits were warned that informers could make a lot of money from the Napa Sportsmen's Club or the San Francisco Club, both of which offered a $50 reward to informers. The clubs could collect the fines from the violators and distribute them directly. (Neither of the clubs owned or leased property in the area, and their right to levy, collect, and dispose of fines is questionable, but legal niceties were not always observed in early California.) (*SR,* 3 Sep 1880)

Whether for hunting or any other reason, trespassing is merely a human-to-human offense. Violation of laws designed to protect game is a different matter. As with pollution, game laws existed in the early days, but there is little evidence of enforcement.

The first effective protection for ducks came not from the state but from private clubs. For the first few years after their founding, the early Suisun Marsh "gentlemen's clubs" shot whatever day of the week their individual members could get away from the office. By the end of the 1880s, however, the clubs were applying restrictions, and in the 1890s, most confined their shooting to Wednesdays, Sundays, and holidays. (Saturdays were only added to the list of open days much later, after the five-day work week became standard.) Local Suisun shooters, however, continued taking time off whenever hunting conditions were right and, as noted in Chapter III, were less than scrupulous in observing property lines. This, in turn, led some of the clubs to open up the other days of the week to shooting so they could compete with the trespassers for the steadily shrinking number of ducks.[2]

Cynics might claim that the practice of shooting two days a week was motivated less by devotion to conservation than by crass economics. Indeed, for money-grubbing businessmen and professionals, it was simply too hard to get enough time free from the office for hunting every day. But it was probably less a matter of conservation or full business schedules than the hard-headed realization that daily shooting would result in lower bags by driving the birds off the marsh. Nevertheless, except when poachers or non-conformist clubs ruined things on the off-days, the ducks profited from the rest given them, and the custom has remained in effect for more than a century. During the no-closed-seasons of the late 1880s, the clubs also imposed on themselves a last-weekend-in-February closing date. (*B & S*,21 Feb 1887)

Legal restrictions followed more slowly, but by 1 March 1891, there were fairly stringent game laws on the books—if not actually in the field. The by-laws of the Sportsmen's Association of California (SAC), the private group we encountered in Chapter III, noted that the penalty for shooting a duck out of season was no less than $100, or 100 days in jail, or both.[3] For shooting white or blue herons the fine could be $50 or 50 days or both. Distribution of the duck fines was 50% to the informer, 25% to the county District Attorney, and 25% to the "Fish Commission Fund for the purchase and distribution of game

---

[2] Starting in the 1890s and down to the present day, there has been a destructive cycle that has repeated itself during poor hunting seasons. With fewer birds to shoot at, hunters go out more often, ignoring the closed-day custom. This stresses the birds even more and leads to still lower bags.

[3] These may have been actually maximum rather than minimum penalties, but they were still extreme for the times.

birds in various California counties." For herons, half the fine went to the informer and half to the "Common School Fund." Two years later the fine for taking herons had risen to $100. (SAC, pp 17, 19, 22, 31-32; F&G Code, 1893)

From 1893 it was no longer lawful to hunt with anything bigger than a 10-gauge shotgun. Although sale of wild ducks continued to be legal until 1917, long after it was outlawed for upland gamebirds (1901), by 1895 the season for *selling* ducks was only from 15 November to 15 January, a month off either end of the shooting season. There were still no wardens authorized for Solano County, but on the 1895 opening day four hunters were arrested on the Suisun Marsh for unspecified violations, probably trespassing. (*B & S,* 18 Mar 1893 and 26 Feb 1895; *SR,* 18 Oct 1895)

Turn of the Century Watershed

There were surprisingly strong currents for more regulation at the dawn of the twentieth century. Through the 1890s, the basic political unit for establishing game laws was still the county, most of which had their own "game and fish" protective associations, but in November 1899, a convention of county delegates met to try to establish a statewide organization. Just a month before, Marin County had passed an ordinance outlawing pump guns,[4] and there was a swelling movement to ban market hunting. (*B & S,* 21 Oct and 25 Nov 1899) These developments were vigorously opposed by market hunters, and a growing split developed between the more progressive bay area counties and those inland. Nevertheless, a statewide game and fish protective association was organized in May 1900 by Governor Henry T. Gage to recommend comprehensive

---

[4] The Marin County fight to ban pump guns continued until 1905, when the Court of Appeals ruled that only state game laws were constitutional. (*B & S,* 5 Aug 1905)

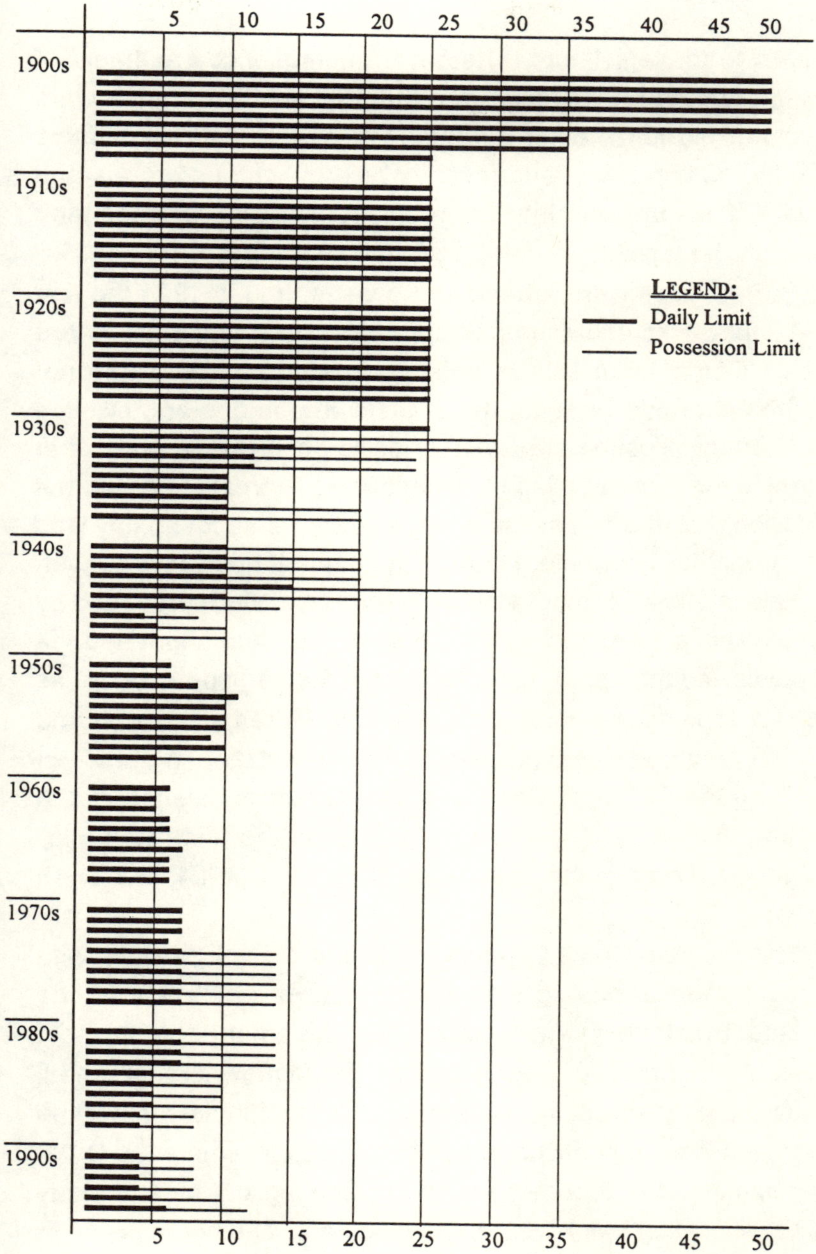

CHART 2 -- DAILY BAG AND POSSESSION LIMITS -- 1901-1995

game laws to the legislature. The government's encouragement of public input was a healthy sign, but the 1890s Suisun range wars (Chapter III) had left a bitter heritage that still divided the hunting community. Given the strong feelings held by various groups and individuals, it is surprising that any regulations at all had made their way through the legislature. (*B & S,* 6 Oct 1900)

But the laws were drafted and passed, and in 1901 the first limits—50 for ducks and doves, 25 for snipe, rails, and quail—were imposed. Commercial sale of upland birds and venison (but not ducks) was banned. In 1903, more rules were added and existing ones tightened. Posting of property had to occur at intervals of at least three to the mile to be legal. The daily limit remained at 50, but a possession limit of 50 was also spelled out. The shooting day was defined as one half hour before sunrise to one half hour after sunset.

Late in 1904, yet more stringent regulations were proposed by a State Sportsmen's Association. These would have shortened the season, outlawed the use of animals or any "floating contrivance" as a screen for approaching game birds, and empowered a county game warden to come on private property without a warrant "in pursuit of his duty." Further, it suggested a $1 resident hunting license, a $10 out-of-state license, and a $25 fee for aliens the last, as frankly acknowledged, aimed at the Japanese. (*SR,* 2 Dec 1904; *B & S*, 18 Feb 1905)

The Fish and Game Commission at first did not go along with these recommendations, preferring the looser approach of the Sacramento (Sportsmen's) Association. On the other hand, they did turn back a move by one state senator to let landowners ignore all hunting regulations on their own property. Interestingly, neither the State nor the Sacramento Sportsmen's Association wanted to lower the daily bag limit to 25, a measure that some legislators and many individuals (including Frank Maskey) advocated. The commission in fact did recommend a 25 bird limit in 1904, but the influence of the two associations prevailed.

The sporting press (*Breeder and Sportsman, Solano Republican,* and *Western Field*), all of which were usually at the forefront of the drive for more and better conservation measures, supported some of these proposals but were decidedly ambivalent on others. The 1905 *Breeder and Sportsman,*for example, noted that "the bag limit is a jest" that hunters "make no pretense of observing," and that lowering it would thus be ineffectual, as well as working a hardship on the law-abiding. The *Solano Republican* went further, arguing that the bag limit should remain at 50 and the season should open a month earlier. But in 1907 the limit did drop first to 35 and then, in 1909, to 25, where it held until the Depression years. (See Chart 2, p. 151.)[5] (*B & S,* 3 Sep 1904, 18 Feb 1905; *SR,* 16 Dec 1904, 4 Oct 1907, 8 Oct 1909)

Also in 1907, counties were instructed to issue the first hunting licenses according to the $1, $10, $25 schedule proposed in 1904 and vigorously opposed at that time by the *Breeder and Sportsman*. (Later, the magazine equally vigorously supported licensing.) The licenses were to be made of "aluminum or other suitable metal," about 2 inches in diameter, and stamped with a number the state would issue for each, together with a June 30 expiration date. The penalty for hunting without a license ranged from $10 to $100, or a county jail term of 10 to 100 days, or both. Proceeds from fines were to be paid into the Game Preservation Fund. But at the end of 1910 it was estimated that only one out of five California hunters bothered to buy a license. Wrote Arend in 1966,

---

[5] Perhaps the most surprising thing about the 1905 game laws was again the severity of the penalties: $25 to $500 fines plus imprisonment for misdemeanor infractions, and for really serious crimes, such as killing an elk (defined as a felony), a mandatory one- to two-year prison term. Using explosives for taking trout cost a minimum $250 fine *plus* imprisonment.

"A hunting license that cost a dollar was available if you really wished to buy one. But the Game and Fish Patrol was more interested in oyster pirates, and a gentleman was not annoyed by inquisitive wardens on his private shooting marsh." Nevertheless, enforcement of other game laws began to be felt, even if the license law was only invoked in connection with other misdemeanors such as over-limit shooting or hunting at night, usually good for fines of $25 to $50. (F&G Code, 1907; *B & S,* 26 Nov 1904, 31 Dec 1910; SSCD.1, p. C-9; *B & S,* 5 Nov 1910)

The dropping of the duck limit to 35 was at the instigation of a private "Fish and Game League" that was proposed by Lt. Governor Alden Anderson in 1906. Anderson, himself a duck hunter and occasionally a Teal Club guest, set up the league and was a member of its legislative committee but left its presidency and administration to others. Long afterwards, in 1945, the California Waterfowl Association was founded with the same goal of letting private hunters influence duck-hunting legislation in Sacramento.

Campaign to Ban Market Hunting

From the time the first limits were imposed, the two state-wide journals most interested in hunting (*The Breeder and Sportsman* and *Western Field*) had taken a firm stand against the sale of wild ducks. The more local *Solano Republican*, on the other hand, said the movement to restrict sales was just a way for the wealthy to reserve all ducks for themselves, denying them to non-hunters. Carried further, the argument said that this would restrict the "common man's diet," a line that the anti-marketers easily refuted by noting that wild ducks were far too expensive for any but the wealthy to buy. The pro-market faction included not only market hunters and their commercial patrons but even some of the clubmen, who boasted of paying for their ammunition by selling their surplus kill. From 1907 onward, *Western Field* carried as part of its masthead a stern enjoinder to the legislature to outlaw all sale of wild game. (*SR,* 20 Jan 1905; *B & S,*

9 Oct 1909)

Western Field made a special point of criticizing the Sacramento duck hunting community, whose Sportsmen's Protective Association had been the foremost opponent of game law reform. Accusing them of being "gentleman game hogs" and "lacking the the simpler ethics of civilization," the editor asked,

> What mercy should be shown men who are so contemptuously bigoted, so maliciously vindictive, so morally irresponsible that they would in a spirit of petty personal spite and revenge repeal all our game laws and so wantonly destroy a public heritage simply because they are not allowed to selfishly batten upon it? Out upon such rascals! (*Western Field,* v. 9 No 6, Jan 1907, p. 901)[6]

Despite the disapproval of the press, market hunting continued. About 1910, however, enforcement of all game laws began to be tightened, and the drive to prohibit market shooting gained momentum. The California State Fish and Game Protective Association, at its 14th annual convention in 1910, had this as one of its main platforms and invited the governor to attend. It was to be a long, bitter fight, however, before ducks at last were protected from market hunters.

Part of the control problem lay in the right of every private sport hunter to sell his surplus ducks. Up to 80 percent of the 28,425 birds reaching the San Francisco market in November 1914 were

---

[6] Ah for the days when editors could really mince opponents instead of just words. But in fact this editor, who was opposed to gentlemen's clubs in general, probably singled out the Sacramento community because it was safely far away. He had to be more circumspect when it came to the Suisun area because one of *Western Field*'s sponsors was the Field and Tule offshoot of the San Francisco Olympic Club.

provided by private hunters. (The total shipment-to-market figure for November the previous year was 40,358, with 480 individuals contributing.) Market hunters supplied San Francisco with a dramatically decreasing number of ducks from 1910 onward, but the figures, though probably only guesstimates, remained impressive, especially if private hunters were not counted in the totals:

1910	500,000
1911	250,000
1912	150,000[7]
1913	85,000

(*B & S,* 15 Nov 1913, 24 Oct 1914; 12 Dec 1914)

Prices for the birds were higher than in the 19th century, but were still not excessive. In 1916, the last year of legal market hunting, prices of ducks per pair were listed as: mallards $1.65 to $2.00, canvasback $1.75, sprig $1.25, teal $.75 to $1.00, widgeon $.85 to $1.00, blackjack (ringneck) $.85, and spoonies $.60. Geese were an even better bargain: pairs of "grays" (whitefronts?) went for $1.15, snows $.90, and brant $1.00. (*B & S,* 21 October 1916)

Once daily bag limits became the law, market hunters theoretically were held to the same limits as sportsmen, but in fact they had to kill many times the limit to make any kind of living, especially after the maximum dropped to 25. At first, the commercial hunters successfully evaded the law by sending each 25-bird limit separately via railway express to a "game transfer" company

---

[7] In October 1912 alone, some 2,272 sacks of 25 ducks (nearly 57,000 birds) were reportedly sold in San Francisco. The take for this period is disproportionately high, however, because it reflects the opening days of the season, when most birds are shot.(*B & S,* 23 Nov 1912)

organized for the purpose. The transfer companies claimed that they were not "possessors" of the birds, which, they said, were private property that they were merely delivering to eventual consumers. But after deputies found the transfer companies sorting out ducks by category for sale to different consumers, the companies' role as commission houses was apparent. In one two-day operation in 1912, the authorities impounded a shipment of 40 sacks containing 1,000 ducks and informed the transfer-company addressees that they could pick them up—but would be arrested if any took more than 25. This put a serious crimp in operations but did not halt them. In 1915, the authorities claimed to have cut commercial sales of ducks in half, but it is hard to say whether that was caused more by law enforcement or duck scarcity. (*B & S,* 23 Dec 1911, 10 Feb 1912, 23 Jan 1915)

The punishment for over-limit possession varied, depending on the culprit, the year, and the judge. In 1913, the manager of a commission house was fined $25 for having 268 ducks in his possession before the season even opened—a punishment of less than $.10 a bird for a double offense—but not everyone fared so well. In December 1916, two market hunters paid $100 each for having slaughtered geese with a quadruple barreled 8-gauge gun fired from behind a bull. (What a kick *that* must have given!) And just a month before, an incautious returnee from the marshes who had asked the help of a Redwood City night watchman in changing a tire, found himself behind bars after the watchman called the warden. The 457 ducks in the man's car were worth $500 and 150 days for the offender. (*B & S,* 8 Nov 1913, 9 Dec 1916, 18 Nov 1916)

Even as the campaign against market hunting gained momentum, a counterforce came into play. When the crackdown on market hunters began, the Fish and Game Commission reported that restaurant and hotel owners had voiced their solid support for stamping out illegal sales. Furthermore, various "game protection associations" had sprung up in California, ranging from the county to the state, and the formation in New York of the American Game

Protective Association heralded a wider conservation effort.[8]

But in late 1914, when the anti-marketing drive began to prove effective, the commercial interests underwent a marked change of heart and were working hard to preserve the old system. At the same time, a new group suddenly appeared on the scene in San Francisco, the "People's Fish and Game Protection Association," whose founder, John F. Corrica, succeeded in getting an initiative put on the ballot that would have removed all game marketing restrictions. With the backing of market hunters and business interests gearing up for the 1915 World's Fair, the initiative got off to a flying start but was shot down by the voters, who felt that Corrica's own police record of 21 arrests for game law violations conflicted somehow with the association's virtuous name. The issue finally died when the sale of wildfowl was outlawed by federal law in 1917. (*B & S,* 10 Oct 1914)

Further Evolution of Game Laws

While the fight over marketing continued, game laws continued to evolve. Limits on geese first appeared in 1915, with 12 for Canada geese (honkers) and 25 for other species. Later, with the signing of the Migratory Bird Treaty Act between Britain, Canada, and the U.S. in 1916, the daily limit on geese dropped to eight (12 for honkers and brant).

In 1915, in place of a possession limit, the law imposed a weekly 50-bird duck limit and a 24-bird honker/brant limit, with the week defined as lasting from dawn Sunday to dawn the next Sunday. No weekly limit for geese other than honkers was written into the code, and it is probably safe to assume that the 8-bird daily limit was

---

[8] These California associations were the forerunners of the California Waterfowl Association, while the American Game Protective Association, in its goals and international aspirations, appears to be the direct ancestor of Ducks Unlimited.

ignored, especially where agriculture was suffering from massive waterfowl depredations. What one might actually *do* with 24 honkers and 50 ducks in those pre-refrigeration days is anyone's guess, though in the 1920s it was still common practice to sell them on the sly and the law be damned. (F & G Code, 1914-1915, 1915-1917; Huber)

At the same time as generous duck and goose limits were the rule, certain species were more protected than they are today. Wood ducks and wild pigeons were protected from 1914 onward, for example. Shooting from power boats was outlawed as early as 1909, as was the use of animals as mobile blinds to approach birds, and the ban was extended to sailboats, aircraft, and automobiles by 1923.[9] Baiting remained legal in those years, however, and in 1910, it was estimated that more than 100 tons of "good wheat and barley" were broadcast for Suisun Marsh ducks each season. (F & G Code, 1909, 1923-1925; *B & S,* 5 Nov 1910)

Starting in 1913, state laws were supposed to conform to more restrictive federal laws, and where the two conflicted, the federal law supposedly prevailed. This meant that the state law allowing shooting a half hour before and after sunrise and sunset, respectively, had to give way to the federal sunrise-to-sunset rule. How much the law was actually obeyed is hard to say.[10] At first, it had few teeth, because Washington's Department of Agriculture was

---

[9] The first recorded instance of a duck being shot from an aircraft in California was by a Frenchman, who killed a bluebill from a flock he pursued out over the surf near Los Angeles in late 1910 or early 1911. His cohorts on the beach confirmed the kill by retrieving it. (*B & S,* 14 Jan 1911)

[10] Judging by today's practices, probably very little. A small card showing sunrise and sunset times was, however, distributed in 1915 as an aid to the law-abiding.

responsible for enforcement, for which it had no budget. At that time, local wardens, who got no federal pay, were asked to enforce the federal regulations. In 1914, however, the transfer of game law responsibiity to the Department of Justice, which did have an enforcement arm, was supposed to remedy the situation, but it took time. In 1916, the difference in the season opener (15 October for the state, 16 October for the feds) was the difference between Sunday and Monday, and when most hunters with weekday obligations chose to observe the state's rule, there were no known arrests. The state wardens did, however, compile an enviable 82 percent conviction record in 1916, higher than in any other class of case. (*B & S,* 13 Oct 1913; *B & S,* 30 Dec 1916)

Only in 1930 did federal law finally and truly become dominant for migratory waterfowl under provisions of the Migratory Bird Treaty, nearly 40 years after it was first proposed. Sharp reductions in the limit and season length followed in 1931. Baiting was prohibited (1932), live decoys were first limited (25, in 1932) and then, in 1935, outlawed. (Oddly, live decoys only became common in about 1907, an indication of how trusting ducks were in the very early days.) The authorities tinkered with shooting times, including noon openers, a practice that continued, off and on, until the 1960s. The last major restriction was the mandatory use of steel shot, phased in from 1983 to 1986, a regulation bitterly contested by Californians and fully imposed only after the federal Fish and Wildlife Service threatened to shut down the whole season unless the state gave in. (PLF, 2 Jun 1987)

Most of these latter developments are within living memory. Bernice Huber, whose father was keeper at the Teal Club in the 1920s, remembers barley for baiting being delivered by the freightcar load to a silo kept on the club just for its storage.) My older brothers recall our father's favorite live decoy, a judas mallard whose dulcet voice was by far the most effective in the flock for pulling in wild cousins. After baiting was prohibited, the large and genial Mr. Black, local warden for the Suisun Marsh, would hold forth in the Arnold

Club living room—his huge hands completely concealing a generous Old Fashioned—on the cruelty of depriving ducks of the sustenance needed for their survival. But in the late 1930s, baiting stopped as the federal wardens took to flying over the marsh (the fruit-drying trays used to suspend the feed were quite obvious from the air).[11] (P. Arnold, G. S. Arnold)

Even in recent times there have been serious suggestions that baiting be made not only legal but mandatory for private clubs. The theory is that the shrinking habitat forces the birds to concentrate on state refuges, where epidemics of fowl cholera or avian botulism can wipe out tens of thousands of birds in short order. By spreading the birds out over a larger area, these epidemics can be at least weakened. Moreover, if all clubs baited more or less equally, none would have an advantage over the others. The feeding would be required for as long as the birds remain in the state—before, after, and during the season. (Powell)

During the 1930s, bag limits were sometimes ignored by otherwise law-abiding hunters at the Arnold Club, who in the interests of statistical accuracy might enter extra birds in the club ledger under names such as F. Roosevelt or H. Hoover. But the temptations for shooting over the limit were becoming more and more theoretical as the duck population plunged. Each club seemed to make its own decision as to which laws one obeyed and which one ignored. The better hunters might still have to call on Roosevelt or Hoover to

---

[11] Baiting in fact did become legal again briefly in 1954 -- but only under California law. The veteran hunter Lew Allen recalls going through the laborious process of complying with all the detailed state regulations, calling in a state warden to verify his observance of the law, getting the latter's approval, -- but then being informed by a federal warden standing at the other warden's elbow that an arrest on federal charges would follow instantly if he so much as dropped even one kernel and shot over it.

explain their bags, but no one at the Arnold Club, for example, ever shot early or late, despite the before-dawn and after-dark gunfire one heard from the neighbors.(P. Arnold)

Nowadays, the early and late shooting seems to be part of the marsh background noise, and baiting is far from being a thing of the past on some clubs. Some hunters use shotguns without the mandatory plugs, which rarely results in more ducks killed but must delight the hearts of ammunition manufacturers (the more shells available, the more haste in emptying the magazine, and the less accurate the aim). Despite continued grumbling, most hunters abide by the law on steel shot. A common and obvious infraction is shooting from a boat under power, which is risky both for one's chances of being caught (the two noises of shotgun fire and outboard motors being easily identifiable and locatable) and from a hunter-safety standpoint. Regrettably, shooting over the limit is also a common sin, though it rarely involves the serious abuses of yesteryear, when hundreds of birds were killed in a "drag."

More Keepers Versus Poachers

As noted in Chapter II, the observance or non-observance of trespass laws in the early days depended mostly on the respective weaponry of the keepers and the violators. The biggest artillery won the argument. Toward the end of the century the law played a bigger role, but not all conflicts reached the courtroom. Neither then nor now have keepers always been sticklers for due process, and many poachers have escalated confrontations to unreasonable levels. Wardens traditionally have had the highest casualty rate of any law enforcement group, but keepers have had to do some of the same job without benefit of a badge or uniform. In the old days, many keepers were deputized as wardens—in 1882, "Cap" Chittenden was even sworn in as a Deputy Sheriff to protect the Cordelia marshes—but their authority was often unconvincing to lawbreakers. (*B & S,* 23 Sep 1882)

In 1912, following the long-established tradition, a Joice Island Club keeper, Jim Graham, took several long-range potshots from an observation tower at an innocent Greek fisherman, Joe De Rosa, who had come ashore on club property. One bullet went through the target's clothing but merely grazed his shoulder. In earlier times that and a few shouted cusswords might have been the end of the matter. Even though Graham was disliked locally for being bad-tempered in general and unusually testy about trespassers, the right to defend territory by whatever means used to be taken as a given. But De Rosa had him arrested, the judge set bail at $2,000, and the paper headlined it "Cowardly Shooting." The Old West was changing. (*SR*, 19 Jan 1912)

In the 1930s, the Arnold Club keeper Gus Davis was patrolling a levee with a little .22 and apprehended a none-too-intelligent poacher. The poacher laughed in his face, pointed his big 12-gauge automatic at Gus, and told him to get the hell out of there. Gus shrugged, walked off about a hundred yards, out of range of the shotgun, and yelled, "Now, you son of a bitch, throw down your gun and those ducks or I'll kill you," and sent a .22 slug whistling past the man's ear to prove his point. (The poacher voiced colorful objections but complied.)

On another occasion, a poacher got the upper hand. He knocked Gus unconscious with his gun barrel and fled before Gus came to. Gus had never seen the man before, but with patient, quiet detective work in his spare time, he finally ran him to earth two years later. The man had made the mistake of boasting about his deed in a bar in Pinole, many miles away. Gus secured a conviction and sent him to jail.

In more modern times keepers rarely have been deputized and have been known to shoot the bottom out of boats left by poachers who arrived by water, or to sabotage motorcycles or ATVs that came by land. This is not a recommended approach. For their part, poachers who resort to threats are usually not too bright. The

Teal's Sam Hartzell once encountered one who waved a .45 automatic pistol in his face. Sam pretended to accept the situation, went home, and called the sheriff. End of problem.

Keeper Vignettes

Keepers themselves are not immune to bad judgment. Gus used to tell the tale of two of them in neighboring clubs who stood guard in their respective watchtowers, occasionally firing off a casual shot from a 30-30 in the other man's general direction and getting a reply in kind—until one day when one fired and there was no answering shot. (Was it the ill-tempered Graham? Or was he the victim? Who knows?)

To make up for occasional judgment lapses, keepers must be tough. In September 1901, an old-time keeper, Otto Hansen, was working at the Tule Belle. He went ashore from his skiff, leaving his dog and a loaded shotgun in the bow. The dog allegedly stepped on the gun, which went off and hit Otto in the left arm. Otto then walked two miles across the marsh to Teal Station, caught a ride on a handcar into Fairfield, and immediately had the arm amputated above the elbow. There was no mention of anesthetic in the operation.[12] (*SR*, 27 Sep 1901)

Keepers tend to have their own ethics, which are not

---

[12] I am always suspicious of gun accidents attributed to dogs. Unable to defend themselves, they are the perfect patsy for hunters too ashamed to admit their own carelessness. But in 1995 there was a witnessed accident on the Suisun Marsh where some damn fool left his gun *deliberately loaded* in a case ("I always do it that way") and put it in the bottom of a boat, where a dog stepped on and discharged it, fortunately blowing a hole only in the boat. Hansen's accident had no witnesses, but I, for one, would not have challenged it—at least not to his face. Even after he had only one arm left.

necessarily shared by everyone. At the Arnold Club, Big Stan, who was born in Victorian times, was scandalized when he learned that the attractive young lady who came along when he hired Gus was not the new keeper's wife. When she learned that marriage as well as cooking would be the price of continued residence, she promptly decamped. Gus went into town, shopped briefly, found a good cook from New Hampshire, married her on the spot, and brought her back. They seemed to be completely neutral toward each other, but the relationship—if there was any—was stable. Mrs. Davis was a great cook, a warm person, and she always looked after the Arnold boys. The Davis couple stayed at the club until it was sold in 1942.

The job of keeper requires not just courage and ingenuity but more integrity and ability than might be apparent on the surface. Though technically only responsible to the club president, many must answer to all members and, in some cases, to their wives as well. A keeper often has either too much company or too little, and it is not surprising that so many of them take to the bottle.

The ideal person is one who has a family with him, doesn't drink, doesn't hunt, and doesn't have friends or clients who hunt. He should be able to maintain and operate the various farm machines that most clubs use to keep their property in shape. He has to be physically fit and mentally alert enough to take care of such sudden emergencies on his own, yet follow directions when the owners are around. In a basically untidy environment, he should keep things cleaned up. Above all, he has to operate ethically when no one is looking over his shoulder, a test that many have failed over the years. In 1910, the duck-hunting police chief of San Francisco, curious why some clubs fared much worse than others (his was one of the former), found that some keepers were leasing blinds to locals on rest days. Several keepers were then reported as "now looking for other jobs." (B *& S,* 10 Dec 1910) That problem still existed in 1983, when Joice Island had to fire its keeper for the same offense, and it will probably always be with us, despite the certainty of instant dismissal for anyone caught in the act.

Most of the keepers I have known have been great people, many of them following in a family tradition. Over at the Joice Island Club, Lee Deming is a third-generation caretaker, and at the Teal, Damian and Georgette Cabral put in a long apprenticeship at the feet of their grandparents-in-law, Sam and Tressie Hartzell, taking over seamlessly when the Hartzells retired. The Montezuma Club's Bob and Joan Hale came new to the profession a dozen years ago but are another of those perfect fits. (How many others there must be—my apologies to those I've missed.)

Wardens, the Law, and Warden-Hunter Relations

The law has done well to enlist citizens' support through such programs as Cal-Tip that encourage reporting of game law violations. (The old *Solano Republican* would have liked that one.) But relations between wardens and basically law-abiding hunters—who should be Cal-Tip's best sources—are often unnecessarily strained. And the ultimate losers if relations are allowed to sour will be what both wardens and hunters want most to preserve: wildlife in general and waterfowl in particular.

At the risk of turning a more-or-less straightforward history into an advocacy tract, I believe that both hunters and wardens should consciously recognize that (1) the *basic* purpose of game laws is (or should be) preservation of the waterfowl resource, not just the arrest of violators, (2) both wardens and hunters share this interest and should work together to promote it, and (3) regulations and their enforcement should reflect the concept more accurately.

This means that wardens and hunters alike should focus on really serious violations: market hunting and other over-limit shooting at the top of the list, and essentially technical violations at the bottom. In between, on a sliding scale, should come penalties for such illicit intentions as baiting and/or illegal capabilities such as unplugged guns or illicit (lead) shot. There should be progressively stiffer punishments for repeat offenders.

Without exactly recommending a return to Robin Hood's times, when poaching one of the King's stags could get a man's neck stretched, it does seem to me that we could use some more attention-getting deterrents. For example, I have found no evidence of anyone killing an elk in California as long as the 1907 law defining that as a felony, good for one to two years in the penitentiary, was on the books.[13]

About thirty years ago, an extremely rich man in Maryland, who kept getting arrested for shooting over baited canvasback waters, got a wakeup call that others might heed. Regularly arrested, he was given progressively stiffer fines that he would promptly and painlessly pay—until the day came when the judge lost patience and zapped him with six months in the pokey. After that, there was no more baiting. The more recent penalty of denying license privileges to particularly blatant offenders is a step in the right direction, but absence of a license is not nearly as stigmatizing as a spell behind bars. Serious offenders deserve nothing less.

On the other hand, punishment for certain technical offenses should be removed or lightened. For example, it does not affect duck survival in any way if a licensed hunter, fearful of going over his waders, has left his documents in his parked car, at the clubhouse, or in some other dry spot where they can be quickly retrieved. Yes, it is against the rules, but the rules should be relaxed or at least left more up to the warden's discretion. I know of several cases where

---

[13] As a case in point, I recently heard about two unlicensed, out-of-season hunters who shot a Grizzly Island elk, bundled it and themselves into a small hatchback Ford (good Lord, how?) and took off for home, the elk's antlers dragging on the highway behind them. They got all the way to Richmond before they were stopped by a curious cop. On second thoughts, it's doubtful that even the threat of a two-year rap would have held much meaning for that level of IQ.

a warden's route on leaving a club passed within a few feet of the proof of innocence of a hunter he had just arrested on a no-license charge, yet he refused even to look at them.

Technical violations will always remain on the books, if only as a last resort for a warden who knows someone is guilty of a more serious offense but cannot prove it. Just as frustrated prosecutors had to fall back on the income tax laws to jail Al Capone, so wardens should have some capability for nailing the too-clever game hog. But they should not overuse their power.

The state and federal wardens who have checked my credentials in recent years have been universally polite, and the service should be congratulated on its etiquette. Their methods of surveillance and control have usually been defensible. The practice of setting up road blocks to intercept hunters on their way home, for example, is an efficient way of checking large numbers of hunters with minimum enforcement personnel, though more care should be taken to avoid needless delays and inconvenience. But the practice of interrupting hunters in their blinds when there is no evidence or even suspicion of wrongdoing is needless provocation that can ruin a day in the field for the hunter and permanently sour his view of the service. If one out of five such checks reveals some minor technical violation, the service may collect one small fine but will probably have lost the potential collaboration of all five hunters. That's a big price to pay.

Back in 1913, when federal wardens first made their appearance in California, they had an operating creed that bears recalling:

> We are more concerned at present in advising the 3,000,000 to 4,000,000 persons who hunt of the law and its provisions and in securing as much support as possible from the 86 or 87 millions of people who do not hunt, than with detecting violations of the law due to ignorance or inadvertence. (*B & S,* 11 Oct 1913)

Since those days, the ratio of hunters to non-hunters has shifted dramatically, but the message remains just as valid.

As hunting license sales decline and government budgets shrink, hunters and wardens are both becoming endangered species. Neither of us can afford to mistrust and antagonize the other. The price of mutual ill-will could be mutual extinction.

✧

CHAPTER 9

BAGS, DOGS, AND YARNS

Bags

Back in the early days, the market hunters said they thought nothing of 250-bird bags, and amateurs were only a bit less boastful. (How comforting it is that hunters no longer lie.) The fact is, though, that even in the days of abundant ducks, hunters scoring in three figures were not common, and then (as now) outside influences affected statistics, both upward and downward.

In January, 1885, for example, when it was rumored that the Teal Club might lose its lease to other rich hunters, the club president not only issued a belligerent statement about his iron-clad rental rights but in the same article also confided that the previous Sunday he had killed only nine birds and the two other club members never even fired a shot. (Message: Not only do I have it locked up, but it isn't worth your while to try to take it away anyway.) In 1887, however, with its lease seemingly *really* secure, the club was boasting that its daily bags for that same 1884-1885 season had ranged from a low of 35¾ to a high of 48¾ per gun per hunting day.[1]

---

[1] The Teal's pride went before a fall when, a few years later, the Cordelia stole the lease anyway. See Chapter III.

(*B & S,* 3 Jan 1885 and 24 Dec 1887)

Nowadays the practice of doctoring figures has not died out just because limits are lower. A fairly sure sign of a club about to come on the market (or already being quietly peddled) is a sudden surge in the reported kill for the season. Any claimed average bag that approaches the theoretical legal limit invites skepticism. The inflated-bag-for-gain ploy is nothing new, either. In 1900, a jump in the reported take of canvasbacks from certain Suisun ponds was laid to the landlord's intention of raising the lease the following year. "The rental is a good one now, but Oliver Twist wants 'more.'" (*B & S,* 1 Dec 1900)

In the 1890s, the highest known one-day take on the Suisun Marsh on three occasions hit 170 and 171 at the Teal and Cordelia clubs. The Ibis reported a total of 16 instances over a 15-year span (1883-1898) where hunters scored 100 or more in a single day. The average for a sporting gun in those times, however, was much lower. A careful tally of the sporting journals' kill figures reported for individual hunters (who usually provided figures only when they had something to brag about, and who rarely underestimated their bags) was less than 25 per gun per day. The press, however, kept citing the figure of 200 birds per day as if it were quite common, and it used the figure as evidence of the preserve hunter's greed, even though no Suisun hunter is known to have scored that many.[2] As in so many other fields, the reported numbers were less a reflection of objective truth than subjective prejudice.

---

[2] When the Tule Belle was still on Sherman Island, to be sure, "several" 200+ bags of widgeon reportedly were made in the early 1880s right after a major levee break, but those must have been exceptional conditions. Mining siltation then dropped the club's total kill from 1,355 in 1883-1884 to 758 the next season, causing the club to move to Suisun in March 1885. (*B & S,* 6 Mar 1886)

Not all the statistics appear to have been shaded. In the very earliest days, the Cordelia Club reported that five of its members had opened the season on 15 September 1882 and secured 55 birds. Granted the season started early that year, and granted that neither the Cordelia marsh nor its neophyte hunters were as well developed as they later became, an 11-bird average was a modest score for an opening day. (*B & S,* 23 Sep 1882)

But the most detailed seasonal bag figure for one season in the 1880s comes from the Tule Belle Club (Table 1). The total kill of 2,324 was gathered in the course of 166 hunter days, indicating an average of exactly 14 birds per gun. But nearly half the species borne home in triumph by yesteryear's gunners would not bring much favor in today's kitchen or game warden's office. If we eliminate the untasty, the illegal, and the borderline (including spoonies, curlew, bufflehead, goldeneye, merganser, and swan), we end with an average of a bit under 9 ducks and geese per gun per day.

On the other hand, over the longer period of 1882 to 1907, the bags of both the Tule Belle and the neighboring Ibis Club showed a marked improvement in quality if not—for the Tule Belle—quantity (Table 2). Quality was refined thanks to the original compiler of these figures, Emerson Stoner, who, in 1937, charitably eliminated an unknown number of herons, pelicans, larks, killdeer, yellowlegs, "robin snipe" (dowitchers), and "Skenk-doo" (bitterns) from the clubs' statistics, from which it can be inferred that hunter sophistication still had a distance to go in the early days. He combined both clubs' counts of swans (71) and geese (488 snows, 73 whitefronts, 34 Canada geese and subspecies) during the period.[3] (Stoner, p. 246)

---

[3] Most of these probably came from the earliest years. Note the fall from grace of the spoony, almost a third of which on the Tule Belle were taken in 1885-1886 alone. In later years, the hunters must have spurned them.

Spoonbill	505	Gadwall	37
Teal	427	Goldeneye	18
Widgeon	372	Redhead	7
Pintail	257	Canada Goose	2
Canvasback	164	"Grey goose" (White front?)	2
"Curlew, etc"	147	Merganser	2
Mallard	106	Swan	2
Bufflehead	89	Wood Duck	2
Jacksnipe	76	Cackling Goose	1
Bluebill	58		
Snow & Ross' Geese	54	**Total**	**2,324**

Table 1—1885-1886 Season Waterfowl Kill at Tule Belle Club
(Source: *Breeder and Sportsman,* 6 Mar 1886, pp. 148-149)

Not even these careful statistics can be accepted as completely accurate, of course. The *San Francisco Examiner's* skepticism about the Cordelia's game log in 1892 could have applied as well to the Tule Belle or Ibis: "The record of those who hunted all day and returned with a seagull, a shag, and two gumboots full of black mud and salt water does not appear on the minutes of the club." (*SFX,* 2 Mar 1892)

Stoner credits the Ibis with an approximate 20-bird hunter/day average for the 25-year span but has no similar figures for the Tule Belle. The latter, however, had at least ten members, whereas the Ibis only had three members for the first five years, and five members thereafter. Assuming that roughly the same percentage of members of each club were on the marsh during the 25 seasons, the average hunter/day take works out to about 5 birds for the Tule Belle, even including the spoonies and "fish ducks" (ruddy ducks, buffleheads, goldeneyes, mergansers) that made up 11 per cent of the take. Note, however, that the percentage of spoonies in the Tule Belle total fell from 22 percent in 1885-1886 to about 9 per cent over the longer period. Whether this was the result of fewer spoonies on the marsh or more discrimination among the hunters is not clear. Perhaps it was a little of each.

Although the Ibis ponds were 3-4 feet deep and those at the Tule Belle only 1-2 feet deep, the percentages of species killed were roughly equivalent except for teal and canvasback. There was no known prejudice against shooting ruddy ducks, but maybe their numbers were fewer in those days. (Later, in the 1930s, they were protected.) The saddest development since those long-ago seasons is the decline of the canvasback, which nowadays is a relative rarity on the marsh except when gales make the bay waters too rough for survival.

Overshooting probably was more of a factor in canvasback decline than with many other ducks. Most of the record bags listed in the old magazines referred specifically to cans, which, when conditions were right, were downright suicidal in their determination to decoy. One example was the 160 cans taken in one morning in 1884 on the Hardland Ponds by two gentlemen named Brooks. One got 91, the other 69. The latter brought his trophies back to his San Francisco lodgings and "spread them out on the floor of a good-sized parlor." The comments of his housekeeper are lost to us. (*B & S,* 12 Feb 1887)

Nostalgia has a wonderful way of inflating bags of bygone years. No doubt about it, the hunting was better overall back then, but if we recall that there were no limits for most of the 25-year span that Stoner covered, a five-bird hunter/day average was nothing to write home about—especially when it included spoonies.

Species	Ibis Club (deep water)		Tule Belle (shallow water)	
	Number	%	Number	%
Pintail (sprig)	10,807	29.9	5,107	24.5
Widgeon	9,920	27.5	4,153	20.0
Canvasback	6,981	19.3	1,308	7.2
Teal	3,415	9.3	6,347	30.5
Mallard	2,094	5.8	701	3.4
Spoonbill	1,209	3.3	1,859	8.9
Bluebill	722	2.0	641	3.0
Ruddy Duck	389	1.6	8	--
Gadwall	200	.6	104	.5
Bufflehead	113	.3	373	1.8
Redhead	62		9	
Wood Duck	--		24	
Goldeneye	8		4	
Merganser	6		6	
TOTAL	36,126		20,844	

Table 2—Comparative Ibis and Tule Belle Club Kills, 1882-1907

(Source: Emerson A. Stoner, "A Record of 25 Years of Wildfowl Shooting on the Suisun Marsh, California", *The Condor*, Nov-Dec 1937, p. 246)

Dogs

"Duck hunting without a retriever is like the play of Hamlet with Hamlet left out," said the *Breeder and Sportsman* in 1892. (*B & S,* 29 Oct 1892)

Most hunters today—and virtually all dog owners— would agree. A good dog increases your take by probably 30 percent and your enjoyment by twice that. (A bad dog?... Well, a bad dog strengthens your vocal chords and increases your vocabulary. It also teaches you patience, how to endure the barbs of your fellow hunters, or, alternatively, how to cope with a nervous breakdown. If there were some way of ensuring that only good ones were let on the marsh, I'd like to make dogs mandatory, not optional—especially if we could mandate only good humans while we're at it.)

Dog stories associated with Suisun hunters were common from the start. Bill Richards had a dog called Squire Dan that went everywhere Bill did, including into the Point Reyes Station bar after a 1903 quail hunt. When the bartender popped a bottle of Napa Soda, the dog instantly "dropped to shot" and looked inquiringly at his master. Or so went the tale. (B&S, 8 July 1905)

Over at the Sprig Farm Club on Grizzly Island, there is an honor roll with the engraved names, life spans, and photographs of the club members' dogs, going back to the original Ace—he has since had a namesake, though in another family—who started his career in 1947. (The human members? Oh yeah, there are a few snapshots of them around, too.)

Ace I was a large, brown, shaggy fellow, one quarter Golden and three quarters Chesapeake, who was a top retriever. Once during the 1949 season, when his boss, Ken Swett, and Jack Keeler were hunting some "navigable" ditches near Pierce Harbor on an ebbing tide, Ace—who had never been known to break—did the unheard-of. Just as a bunch of sprig were working into range, he sprang from the boat, flaring the birds, and disappeared over the nearest levee, five feet above them.

Moments later, Ace silenced his owner's cursing by reappearing with a wounded bird, a fine cock sprig. No sooner had he delivered it than he went up and over again, and again returned with a live sprig. This repeated itself until he came back with his 11th and last cripple, a little teal. The hunters had not fired once, but now had one over a double limit. Ken quietly stuffed the teal in his jacket's hand-warming pocket, and, with the the other birds proudly displayed on the decked-over bow, putt-putted back to Pierce Harbor.

It had been a very slow season, and the sight of a double limit of sprig caused a sensation among the hunters gathered there. The boys stonewalled all queries until the little teal provided one last bit of excitement. Suddenly recovering, he poked his head out of the pocket, took one look at the crowd gathering around, and jetted into the sky, away to freedom.

"What was that?" asked one onlooker.

"Rats," said Ken, "Best damn live decoy I ever had—and now he's got away!"

Ace usually sat on the front seat of Ken's pickup in regal majesty. He had no objection when someone would put a hunting hat on him, and strangers occasionally mistook the silent, motionless passenger for an unsociable human. For all his fine qualities as a retriever, it was probably more his personality than his effectiveness that led Sprig Farm to start honoring their dogs. What a great tradition!

Ray Lewis tells the tale of viewing a neighboring club through a powerful telescope and witnessing a scene that was the funnier for being completely without sound. One of his friends was out with his son and a new lab puppy on a still afternoon. A wayward spoony came over, the father sailed it, and the dog promptly bolted to retrieve—in the opposite direction. The father's cheeks distended as he blew on his whistle, waved his arms, and finally turned and gestured to his son, who climbed out of the blind and finally chased the dog down. Dragging him all the way back to the fallen bird, the son shook his finger at the dog, pointed down, and finally got the dog to pick up the bird. He then turned his back and marched back toward the blind. The dog followed for a short distance, then dropped the bird, and lifted his leg on it. The father's outrage, again in silence to the viewer, was communicated to the son, who turned back and dragged both dog and duck to the blind. That evening Ray met the man at a party. "How's that new pup of yours doing?" he asked. "Oh, pretty good—yes, very good, in fact. He still has a few things to learn, I guess, but he's actually great!" "Well, that's good—Say, how did you teach him to pee on spoonies?" And the guy looked as if he'd seen a ghost.

My dog, Centerfold, is quite fastidious about spoonies. Whenever I send her out to get one, she starts with enthusiasm, but as soon as she gets to the bird, she has her own pantomime. She looks down, her brow furrows, and she looks back at me. "Oh no, you didn't... you *couldn't* have done that," she says, and brings back the bird with the air of someone carrying a dead rat.

Mind you, there are times when the human retrieve is more reliable than the canine. During a dense fog many years ago, a gunner in one blind heard a shot from his neighbor, followed by an excited "Fetch" command to a dog, and a great splashing. Just about then, a mortally wounded mallard sailed out of the mist and collapsed, dead, close to the blind, as the splashing dog drew nearer. Thinking fast, the gunner plucked from his own string a drake spoony and threw it in the direction of the approaching dog, who seized it and bore it back in triumph to his master, whose gratitude was expressed in terms unsuited to this family book.

Yarns

It is unkind to suggest that a classic duck hunter could ever tell a lie, but the following from Bill Richards might have stretched the truth a little. Back in the 1890s, when Bill was in New York, some miscreant stole his shoes. The police found robber and shoes, and, under the impression that Bill was also missing his socks, returned to Bill those they found on the suspect. The latter, it turned out, was a deaf mute, and he raised vehement but unintelligible protests when relieved of the footwear, eventually dropping dead of sheer vexation. On receiving his own shoes and the other's socks at his hotel, Richards noticed an unusual stiffness in one sock, which turned out to have 10 $1,000 bills sewn into it, obviously the loot from some earlier robbery. Unable to trace the original owner, Bill had no recourse but to keep the money. (The reporter for this story seems to have accepted it as gospel.) (*WF*, 1907, pp. 854-855)

Despite the ever tighter game laws and enforcement noted in the last chapter, an outlaw fringe persisted and, unfortunately, survives to the present day. I know one man (Mr. X, who is still alive and therefore must remain nameless), who was a flat-out market hunter and only stopped when old age and modern enforcement technology caught up with him. He used to provide the old Matson Line with ducks that were served up after the passenger liners passed beyond the 12-mile limit on their way to Hawaii, and he later supplied ducks to at least one San Francisco restaurant. He also, I suspect, often broke the law for the sheer joy of outwitting the wardens.

Mr. X's techniques for skirting the law were many and varied. When a big restaurant order for ducks came in, his first step often would be to have the owner take out licenses in the names of all the restaurant staff and forward the license numbers to Mr. X, who would recruit local boys to hunt his property. Once they had limits of good birds, Mr. X would tag them with the various dishwashers' and waiters' names and license numbers and ship them down to the restaurant. The local boys would then be given a chance to go out and hunt under their own licenses for a second limit. (Mr. X thought he invented that one, but it was just a variation on the old "transfer company.")

He had other tricks, some of them impromptu. On one occasion, having been hunting with a state senator who was quite nervous about breaking the law, he decided to avoid a possible road block along his exit road from Grizzly Island by taking his guest to and from his club by boat, from Pierce Harbor. But on their return with far too many ducks, they noticed that a familiar green car was waiting up where the road from Pierce Harbor T'd at Parish Road.

"Not to worry," said X, "Here, you take all the birds and do as I say."

"M-m-me?" asked the senator.

"Yup," said X, "You just wait till I leave, then you go, and turn left on Parish."

"But hold on..."

Too late. X had seen that the warden was busy checking another hunter, so jumped in his car, started down the road, and when he came close to the warden stepped on the gas, shot around him with screeching rubber, turned right, and floored it. The warden dropped interest in the hunter he was checking, jumped in his own car, and finally, with siren on and red light flashing, caught up with X a couple of miles down the road. Meanwhile the senator had quietly turned left on Parish as instructed and continued home, carefully observing the speed limit.

A far less game-hoggish lawbreaker was Old Man Hugg, an elderly poacher who used to live at Pierce Harbor in a trailer until his death a few years back. When he was not sneaking ducks on other people's property, Hugg would do such good deeds as bailing out open boats that were about to sink from heavy rains or, in one case when my young son fell off the dock in a chilly dawn, warming people up in his snug quarters. Nobody ever asked him for such services—he just did them—and he never took anything in return. One time when I asked him if he wanted some ducks, he answered, "Naw, I've got 14 (the possession limit in those days) sprig and mallard in my freezer. In fact some of 'em came from your club." (He could have had all he could shoot, as far as I was concerned.)

At the far southern extremity of Grizzly Island there was a hut perched in the bayshore tules that we in the Tip End Club knew unaffectionately as the Poachers' Shack. From the rear it looked grim—a patchwork of dark driftwood with no doors or windows that faced in our direction, a jumble of mismatched parts that mostly concealed the makeshift dock that gave the squatter-owners access. It was mysterious, sinister—a place for dark deeds. All we knew for sure about the owners was that they hunted the offshore boat blinds in Grizzly Bay, but we strongly suspected that they poached our land, and we heard rumors that they were locked in a Mafia-style feud with another set of offshore shooters from the other side of the bay. This was serious business, went the story, and best you not be carrying a shovel for any reason when you went near their shack. Never could tell what—or who—might be under the bay mud.

So we didn't go near it at all, despite occasional suggestions from hotheaded Tip End members that we burn them out. Our more peaceable members would point out that we were equally vulnerable to arson, and that would temporarily still the voices of violence until the whole affair would come up for discussion the next year. (Duck club meetings often have re-runs of old topics.)

After our club was evicted from Tip End, I got to know Allan Steinau, occasionally a guest shooter at the Poachers'. He was able to confirm most of what we'd heard. He himself had trespassed on our land and, at one point thinking he was being pursued by our keeper, had hidden in a patch of greasewood. Lying on his back and looking skyward, Allan had been faced with a critical dilemma as a big drake mallard came up into the wind and hovered over him. Discretion won out, and he didn't shoot. But twenty years later the regret still hangs heavy on him.[4]

Regarding the feud, Allan said it was between two rival Italian families, the DiMaggios from the east side of the bay and the Capellas from the west side. An invisible line ran down the middle of the bay, and woe betide any member of one clan who crossed into the other's territory. The feud was bitter, and Allan was present when its last episode unfolded. Sandwiched between shrieks of wind one stormy night, they heard an odd bumping noise coming from the dock. Somebody peered out and discovered two huddled figures trying to start a balky outboard. It took some persuading, but finally two cold, wet, and frightened hunters responded to the members' insistence that they come in and warm up. They huddled off in a corner, politely but resolutely refusing any food or drink, until one of the members said, "Hey, you guys are DiMaggios, aren't you?"

Their looks of dumb horror confirmed the fact.

"Hey, what ever became of Old Man DiMaggio?"

"He died."

"He *did*?! How long ago?"

"'Bout two years. Uh, er, How is Mr. Capella these days?"

"Aw, he's dead too."

"He *is*?! When?"

"Three years back." And at that point the feud died too.

In the fall of 1995, through a series of flukes, I got an invitation to hunt out of the poachers' shack. It was amazing the difference between the back and the front. A picture window looked out over the bay from a bright, tidy room decorated with photos and stuffed

---

[4] Actually, Tip End was far too lowscale to hire a keeper. Perhaps it was a warden or, more likely, another poacher.

birds. A cheerful table cloth lay under one of the best Italian meals I've eaten. The clubhouse was several cuts up from Tip End's, whose main features had been dirt, black widow spiders, and wasps. The club members I hunted with were a couple of consulting engineers whose only language was English and whose hunting ethics were my own. Seldom have I been in a more compatible group.

But I made sure not to bring a shovel.

In the 1970s, the new Teal Club owners and two guests gathered for a convivial evening. Eric Schou, of Norwegian background, thought to introduce the company to the joys of Scandinavian *akvavit.* This is a beverage that seems more and more innocent as it gets colder. Eric's had been in a deep freeze for some time.

After the usual cocktail hour and wine with dinner, Eric broke out the *akvavit*, which he served in glasses that had no stable bottom—you couldn't put them down until they were empty.

Poker cards appeared. The two guests were both supposed to shoot next day at the Jacksnipe Club, a mile away along the SP tracks, and eventually one of them decided he had to leave. Not having his own car, he borrowed one from a member and disappeared into the night.

Time passed. When the cards became illegible, it was decided that the remaining guest, whose name was J. O., would be shepherded by the others to the Jacksnipe in his own car (actually, his wife's new Buick) to retrieve the Teal member's borrowed vehicle. Outside, they were confronted with a super-dense tule fog, through which the intrepid crew navigated carefully and successfully until they had to make a right turn into the Jacksnipe parking area after a grade crossing. This they did a bit prematurely. When the car didn't move all that well, someone got out and discovered that they were high-crowned on the tracks.

No problem. The Jacksnipe had a jeep that could doubtless extract the Buick. Unfortunately, no one could find a chain or rope, but perhaps a push from behind could get the stranded car over the rails? This was tried, but the only effect was to put a slight dent in the trunk. J. O. was horrified. "My God, look what you've done to my wife's car. She'll *slay* me." He went on at some length.

And then the rails began to sing.

At that point, according to one participant, the crew did the only intelligent thing of the evening. They staggered *toward* the oncoming freight, hoping to flag it down with a feeble two-cell flashlight. They hadn't made much progress when it became clear they had better get off the tracks, which they did, falling into the flooded marsh along the right of way. As the oblivious freight thundered past, they rose from the mud just in time to see the Buick's cataclysmic death light up the fog. On the far side, the pieces rained down for some time while the train's brakes howled.

Lost in admiration, up to his waist in the marsh, J. O. had only one comment: "Wow," he said, "did *you* see *that*!?"

Next morning, the three Teal Club members awoke painfully, having made their way home on foot along the rails, but having been unable to convince J.O. that this was the only sensible course to take. With a misplaced sense of legal obligation, he had stayed behind to deliver his report to the police. In mid-morning, it was a sorry crew that made its way into Fairfield to bail J.O. out of his cell, but they were too late. As they drove into the parking lot, here came J.O.'s own car, his wife at the wheel, her husband beside her. Untouched by soap, water, or razor, J.O. was peering through a mask of marsh mud with eyes like overheated bearings and the unmistakable air of someone who had already heard a few more things than his fragile condition warranted. As they met, both cars slowed, and the windows rolled down on the drivers' sides.

"Cute, isn't he?" spat out J.O.'s soon-to-be-ex-wife, and, clashing the gears, she laid rubber in a violent departure.

The story does have a happy ending, however. J.O.'s Browning, which was in the Buick's trunk, miraculously survived, though badly dented. He had it rebuilt, and it has served him well for the intervening two decades.

Not *all* marsh stories can be accepted as gospel, of course. The following is from *The Breeder and Sportsman* of 8 December 1898:

> Bart Wyman is rapidly becoming, under Secretary Markland's tuition, one of the crack experts of the Black Jack Club. The secretary has recently applied for a patent on a device for making flock shots. The Invention looks like a churn. A powerful spring inside the flock gun hurtles a large net through the air a long distance and with immense velocity. The net, as it leaves the catapult, opens automatically and whirls around centrifugally. A series of duck-calls arranged about the margin are set to work calling; this will cause a flock of ducks to fly right into the net which then collapses and by means of a cunningly devised electric apparatus returns loaded with ducks to the exact spot from which it started.

(I don't know—maybe it's true. In fact, to judge from the calling we hear next door, I'd almost bet they're using one.)

**Centerfold (dressed for a day in the field)
and two faithful companions**

Drawing by Roland Clark

"Wings of the Morning"

CHAPTER 10

EPILOGUE
IS THERE A FUTURE?

Is there a future for duck hunting on the Suisun Marsh? I think so, despite one of the most discouraging conversations I can ever remember having.

It was on August 4, 1994 that I interviewed Dan Chapin, one of our era's major heroes in the ongoing fight to save hunting and wetlands in California. Dan was retiring from CWA and moving to Colorado, and I wanted to know why.

He rambled on for a while about having looked in South Carolina, the Ozarks, Oregon, Washington... but Fort Collins reminded him of Sacramento 20 years ago: nice rural city with enough intellectual college life around to keep your interest up—and Denver only an hour away if you wanted high-class culture. Good fishing. Besides, his wife Wanda's asthma...

I asked, how come not the California north coast? No smog, close to San Francisco, also good fishing, and no snow to fight...

And then came the truth, crystallized by a recent business lunch with Governor Wilson and four high-power executives. Dan was the only "enviro-nut" (his term) present, and he had suddenly realized that the environment issue in California was being totally eclipsed by illegal immigration, joblessness, increasing crime, lowering of work-force qualifications and education standards, the shrinking tax base, and the flight of industry and high-powered executives to other states...

"Like you," I interrupted, seizing on the last item.

"Well, yes. But I can't stand to see my 25-30 years of work crash because of these other problems."

"Come on, you laid the legal base for fighting the good fight; we're holding our own because of that."

"Nothing can avoid change. To stand still means to lose ground. As a young man, now, I'd welcome the challenge, but not today..." and he tailed off into sad silence, obviously unsure that any young man, however smart and dedicated, could take on these challenges successfully.

I came away profoundly depressed, especially after I'd heard similar words from an old Fish and Game friend who was dubious about introducing his children to hunting because he feared the sport would die forever just in time to break their hearts. This book was then in its earliest gestation. Would it end up being just a tombstone to the Suisun Marsh?

And then I began digging into the past, and the more I dug, the more I became confident that, despite Dan's words and other evil portents, we somehow are going to make it. The reason for this confidence, paradoxically, is the almost unbroken succession of gloomy predictions that I found stretching all the way back to the beginning of organized hunting on the marsh. Here is a sampling.

1882—"The shooting for water fowl has not been this year what it formerly was.... No doubt the gradual increase of population and the setting up of localities [settlements] near the former feeding grounds of the birds has much to do with it. In the Suisun marshes, ... the sport has been so poor that the shooters are discouraged. They will have to be satisfied that they have had a good day's shoot if they kill half what they used to." (*B & S,* 30 Dec 1882)

1900— "Members of gun clubs... are very reluctant to mention the results of a shooting trip to anyone but intimate friends....This feeling has been promoted by the actions and talk of certain game protection busybodies whose views are so radical and extreme that a strong feeling of disgust and distrust has been the result." (*B & S,* 17 Nov 1900)

1912—"The Suisun Marsh... has failed to live up to its opening promise.... Excessive baiting and the closest care have failed to hold the birds and the shooting is now showing the uncertainty that precedes their utter disappearance." (*Western Field,* Nov 1912, p. 13)

1915— "Not many years will pass before the hunting as it is now known will pass from these marshes, and it behooves those who love the sport to make the most of the present." (*Solano Republican,* 15 Oct 1915)

1916—"Less than fifty percent wildfowl was killed this season than last.... (I)t is a matter of record that less ducks have been killed as each year passed, and it may indicate that ducks are rapidly decreasing in numbers." (*B & S,* 5 Feb 1916)

1948— "In the event that duck hunting shall be prohibited by law, Lessees shall have the right to cancel this lease agreement." (Poopdeck lease)

1950—Mat Keller's father makes special effort to take him and his brother hunting on Grizzly Island because the end of hunting is in sight. (Keller inteview)

1966—"The Suisun club hunter views the future dismally. He sees the Suisun Marsh urbanized, industrialized, or militarized. He hopes it 'won't happen until he gets too old to hunt ducks,' but he wouldn't bet on it." (Philip Arend, *Final Report, Phase I,* Suisun Soil Conservation District, August 1966, p. C-33.)

All of the above sources have gone to the great marshes in the sky. Do their failed predictions mean that Dan Chapin will certainly be proven just as wrong as they? Of course not. The problems we face today are all the ones our forebears had and a few more. It is not only the increasing population and development pressures, not only the "anti's" and their misguided propaganda, not only the encumbering bureaucracies and overlapping jurisdictions, not only the cycle of droughts that have led until recently to lower limits and shorter seasons, but some new and less familiar threats.

Here are some of them:

1. The modern world's easier and more affordable amusements than hunting. Youth today has far more options for fun than there were when today's older generation was growing up.

2. Increasing problem of finding good, free hunting as the easy-going days of just going out on a public marsh or asking a landowner's permission were all one needed to get a day's sport. Thanks to the opportunities the state-managed refuges still provide, and thanks to the open waters of the bays and sloughs, hunting is not entirely restricted to the fat cats, but it is getting progressively harder for the man of limited means.

3. The result of these factors is an aging hunting population, one that is losing in numbers and therefore in political clout, even though it has considerable wealth and has demonstrated its willingness to use that wealth in support of both hunting and conservation.

Why, then, my optimism? For one thing, in similar societies abroad, especially in Western Europe, hunting has survived despite over-population, over-regulation, and other pressures that make ours look easy. Perhaps it is no longer true, but only about twenty years ago, someone calculated that if you crammed all the people on earth into the United States, you still would not have the population density of England. Yet hunting survives in Britain, and it is not restricted to the wealthy. Since 1980, winners of a raffle put on by a British hunting/conservation organization have been coming to visit here in California, and every one of them has been somewhere in the economic middle class, ranging from the moderately wealthy to a heavy equipment operator. (Every one of them, incidentally, has also been a good shot and a fine hunting companion.) If hunting can survive as a sport for the average man in England, it can survive anywhere.

For another thing, there is now a better understanding here between the hunting community and the purely nature-conservation groups than there used to be. I am not speaking here of animal-rightists, whose emotions immunize them against logic and prevent any meaningful discourse, but about groups such as the Audubon Society, the Nature Conservancy, and others who recognize the tremendous financial and political contribution that hunters make to the preservation of wildlife. Obviously, we don't all see eye to eye on all issues, but to the extent that we can find common ground, we are both the stronger for standing on it, and that is now recognized.

We seem to be at the start of another of those long cycles that see increased waterfowl populations and a corresponding upswing in hunter interest. The last one we had lasted for a decade, between 1973 and 1983, to be followed by a decade of lean years, but the signs now look good again.

And finally, despite our aging community, we do have younger people in other parts of the state. I think of Rich Radigonda, who, on a volunteer basis, is teaching waterfowl art and some basic hunting truths to grade school children. I see other volunteer leaders who raise money for CWA, organize bird banding activities, contribute skills and materials for wood duck nesting boxes, and pitch in on a hundred other projects. I see an Iva Rogers mobilizing volunteers up north to handle avian cholera and botulism outbreaks.

Let's not forget that on the Suisun Marsh the Bancrofts, Chapins, Coons, and Frosts of the last generation took on some financially and politically heavyweight opponents—and beat them. All four of these leaders are now in full or semi-retirement from the fray, but on their heels came an overlapping new generation.

Over the past two decades the most active—and feisty—scientific and technical advisor on waterfowl habitat maintenance and improvement has been Paul Crapuchettes, owner of the New Family Club. Retired from a successful electronics engineering career, Paul became a fulltime marsh resident at his club in 1977 and hit the ground running as an advocate of environmental management. He still hasn't slowed down. A tireless hands-on experimenter, communicator, and defender in marsh matters, he never misses a controversy and can be counted on to raise one if no one else volunteers.

If Paul has been the technical guru, Lee Lehman as SRCD's chairman has provided the essential guidance through the bureaucratic underbrush that surrounds landowners and grows taller year by year. Lee did not come by this expertise by training—he was a contractor on the marsh before he left that to work fulltime on SRCD matters—but he was forced to learn it on the job.

Lee's fellow SRCD board members (in 1996 Tim Egan, Chuck Morehouse, Greg Palamountain, Rob Radoff, and George Tillotson—all duck club property owners) take time off from their busy careers to study and advise on developments on and around the marsh. Their monthly meetings, open to the public, traditionally bring together representatives from Fish and Game, the California Department of Water Resources, the California Waterfowl Association, and other interested parties.

The sad thing is that to many hunters on the marsh these names are known either dimly or not at all. Just as in the past, the future is almost certain to bring new formidable attacks on the marsh from unexpected quarters, some possibly even from people masquerading as allies, and stopping them will call for energy, intelligence and dedication. The present SRCD administration has done a superb job in guarding the marsh's interests, but they, too, in time must step down in favor of younger blood.

Will the new Light Brigade now please step forward?

Appendix A

Perceptions

For most people, the Suisun Marsh is no beauty: she is usually too hot or too cold; she is also dirty, drab, windy, buggy, clammy, smelly, fickle, and treacherous. Scarcely good mistress material.

Then why do we love her? God only knows, but the late Philip H. Arend, after running a survey of landowners there some 30 years ago, summed it up well: "One opinion was unanimous. The club hunter of the Suisun has a deeply emotional attachment to this marsh. He wants to see it remain as duck country. He wants his children to have the ineffable thrill of squatting in a soggy blind on a drizzly dawn and taking a pass at a lofting sprig." The last was his way of saying that you can't explain love to those who don't feel it. (SSCD-1, p. C-32)

Being less intelligent than Arend, I will try.

First of all, her voice. In a quiet dawn there is a brooding quality that makes even the old and deaf (maybe particularly them) strain their ears. Especially with an overcast to dull the eastern sky, it is the perfect staging for that most lonesome of sounds, the bugling call of a single high goose. (The opening bars of Brahms' violin concerto, deep and strengthening like the dawn itself, fit this scene perfectly—or would if he had programmed into its meaningful pauses that faint and fading call.)

Young ears can pick up the hiss of set wings and the swish of a landing bird when it is still too dark to see. They hear the soft quacks and whistles that for me are hidden in a perpetual ringing, the legacy of too much bygone gunfire. But even I, by cupping my ears, can pick up the muted gabble and brief, hysterical laughter of the cocktail parties that go on in the marsh when the ducks are in.

There are lots of other instruments in the marsh symphony. Before the clapper rail became rare (and my ears gave out), it gave a descending "Eep-eep-eep" that sounded like a scornful giggle when you shot and missed. On a still morning a wren, perched at impossible angles and busily rustling in the undergrowth around the blind, is surprisingly loud. Redwing blackbirds have a joyous chorus that never fails to cheer ("Most carefree noise in the world," says my wife, Ruth), and an October night roars with lovesick grasshoppers fiddling to prospective mates.

Nowadays there are some unfortunate discords too: the massive C-5As that scream in heavy slow motion to and from Travis Air Force Base; the constant bicker and snarl of traffic along Route 680 that you hear when the west wind blows; the frantic squawks of the so-called power-callers, people who think they can attract ducks by nonstop noise. At such times I yearn for the clout wielded by hunters only a few decades ago and idly dream of making CalTrans and the Air Force shut down their respective operations for the duration of the season. But the power callers would probably be beyond my power to abate, being by definition blowhards and thus having an advantage in the influence games men play; hell, maybe they already *are* the closest thing we have to yesterday's duck-hunting bigshots.

Oddly enough, I'd leave the SP and the thundering transcontinental trains alone. They were there when I was a boy, so of course they are an acceptable, indeed a necessary, part of the marsh symphony. But missing now are the deep and musical notes of yesterday's steam locomotives that were so much a part of it. Their warnings to the grade crossings ahead were in the standard two-longs/one-short/one-long pattern, often starting with a second or so of contemplative rumbling before crescendoing into a great bellow of alarm, then fading in a descending wail. (A great sound, now vanished except in nostalgic railroad museums, to be replaced by the blatting honks of today's diesel-electric air horns, with all the romance of a Mack truck.)

On a wild morning, of course, the wind makes its own music, drowning out all other natural and man-made sounds. The whole world is one vast, indescribable noise, which eases matters for the deaf, who no longer must worry about such delicate fine-tuning as approaching wings. Eyes, not ears, are what count, and it is no time to contemplate the notes of a Brahms or even a C-5A.

To be fair about audio perceptions, we should also look at things from the waterfowl's viewpoint. In 1911, before the horseless carriage was quite so common, there were complaints about the effect of cars on hunting: “About 4 o'clock in the morning...the chug-chugging of the machines and glare of the lamps frightened ducks in thousands from the ponds, where they were resting. When the hour for shooting arrived, the birds were gone.” (*B & S,* 21 Nov 1911)

Then, in the early 1930s, the appearance of a puttering biplane over the marsh would produce mass hysteria among the geese on Joice Island. They would spill into the air in screaming masses, apparently convinced that the Godzilla of all marsh hawks was after them. (And according to Old Sam, our retired keeper, steel-propellered planes near San Jose in the old days indeed would buzz agricultural land to flush and chew up geese that were feeding on crops; I'm not sure about that one.)

Visually the marsh, for all her drab colors, has a beauty that changes constantly. Oddly, I believe the myopic—that's me—in some ways get more out of what they see than do the keen-eyed.

Standing guard over the whole marsh is Mt. Diablo, a looming presence twenty miles off to the south. In the right sunlight, the low hills at its base can be startlingly clear and golden, but the mountain's upper parts are always black, which may explain the Patwin Indians' belief that it was home to evil spirits. Diablo is not grandly sinister, however. It just squats there, with dour solidity. Nevertheless, it is the beacon that gives you your bearings when the marsh turns you around, as it often does, and over time I have come to feel more and more respect and affection for it. From my home in Novato, from Interstate 80, from the East Bay, even from Route 49 in the Sierra foothills, it is easily visible on a clear day, reminding me with its grumpy, ugly profile that the marsh is still there, waiting.

As a boy, I didn't realize I was nearsighted until fairly late in the game, well after my first exposure to the marsh at age 12. On our way out to the blind by boat, there would often be a red streak of dawn in the east, a streak that would be blotted out by the thousands of birds scared up by us and others, and leaving now for the Joice Island refuge or the open waters of the bay. I could not see the individuals unless they jumped from the levee next to us, but the swift eclipse of the eastern sky never failed to awe me as did the roar of their departure over the buzz of the outboard.

Once the sun had risen, the marsh became a mysterious infinity of pickleweed, tules, and water stretching blurrily eastward to the horizon (and extending on to a thousand horizons beyond, for all I knew). Westward lay the soft-textured, sensuous rolls of the Coast Range, also so seemingly far away that I imagined it would take days to reach dry land if we had been marooned there—a fond fantasy of mine at the time. The marsh's magic dimmed a little when eyeglasses let me see clearly at least some of her far-off boundaries.

I still use some of the tricks learned before specs cleared things up for me. For example, when ducks disappear by dipping below a dark horizon, you can often keep track of them by watching their reflections in the water. The same system works if they have caught you out of the blind; looking down at the water to mark their progress is far better than looking up and flaring them with your bald face. I'll never forget one calm day when a big greenhead worked in while I was trying to rearrange decoys for my brother Stan. I hunched over and watched his reflection grow and grow as he bored in, apparently attracted by the ripples I'd made. When Stan dropped him, he was only about 20 yards up and apparently convinced he'd found in me the biggest hen in the marsh.

Weather affects vision as much as it does hearing, touch, and scent. If there is a strong wind out of the north, the sky is usually bright and the hunting good. The air has a clarity that lets even the nearsighted think they can see, and birds can work at any time of day. Northers usually hit two or three times in a season, sometimes lasting for several days, and when that happens the duck-hunting community cancels all other appointments and obligations. There is a sudden sharp rise in workers' claims of family emergencies, in students' urges to do solitary research, and in executives' compelling needs to meet confidential contacts.

By contrast a rainy morning, with or without a wind, often means little or no duck action. There are exceptions; the very beginning of a storm can give wonderful results, but at most only for the first couple of dawn hours. After that, the ducks simply disappear into some other dimension that no man has ever found. You occasionally see high bunches running before a storm wind, or fighting their way against it yard by painful yard, or paying off before it in defeat, but such bunches are rare, and it is unusual for any of them to work into the marsh. Some say the birds spread out like an oil slick into new, rain-created ponds, but it happens too fast for that; the ponds don't form until long after the birds have vanished.

It may be that storms make ducks want to fly high and far just for the hell of it. We know that sprig can make a 150-mile circuit for no apparent reason, returning to their starting points after some hours without having touched down at all. I can understand that. If I had a sprig's wings I'd do it too.

Few things are as enviable for sheer beauty as a duck on set wings. Powered flight is all very well, but it looks like—and is—work. Planing down on the strength of gravity alone just has to be the most wonderful kind of show-off fun, the same kind of feeling that motivates a human springboard diver, but infinitely more complex to achieve. A pilot whose aircraft has suddenly lost power and who must come in dead-stick may, perhaps, get a feeling for the problems involved, but he gets precious little joy out of the clumsy maneuvers that are forced on him. The duck's instantly variable wing is far superior to anything devised by man, and he uses his speed to flare out and even rise while he is still coasting.

Whether it is mallards against the last glow of sunset, sprig at high noon, or widgeon suddenly on top of you in a fog, the confident, effortless grace of a gliding duck has no parallel. And my very favorite for sheer hell-for-leather speed and reckless abandon as he tucks his wings into maximum sweep-back and dives on the marsh is the lowly spoony.

Rain makes real problems for the nearsighted who are dependent on glasses. Contact lenses provide a solution for some hunters, but if you are also astigmatic they don't work too well. In a rain, eyeglasses fog and weep, eventually reaching a crossroads where you can see better without them than with. And at whatever level of correction, even the sharpest-eyed human is a Mr. Magoo compared to any duck, whose vision is simply a quantum jump ahead of ours.

As for me, I love to be out on the marsh in a rain, even if the world is a blur. For one thing, you are sure to have her to yourself. For another, the grandeur of a big storm, with its plays of light and shadow and slanting rain squalls, doesn't require specs to appreciate. And somehow no amount of experience can convince me that a storm won't provide shooting. That's the way it was on the East

Coast, where I spent some hard-rock seasons scratching out only a few birds each year, and if Suisun birds don't know better, I don't really care—I'll go out anyway. (For perverse and peculiar individuality, it's hard to beat the dedicated hunter.)

Illusions on the marsh range from the prosaic to the mysterious. At one end is the marsh's seeming solidity when it has been drying all summer. In many clubs it is easy to drive a car out onto the flats to inspect blinds or perform other maintenance without slogging through the grasses on foot. Newcomers with four-wheel drive, especially, are prone to breezy overconfidence. A damp spot in their path has no more significance there than it would on an Interstate—until they try to go through it. On two occasions in the past decade I have seen large Suburbans all but vanish in the liquid peat that lay waiting just inches below the surface; in both cases it took the better part of a day's work to get them out. Even the dusty surface of salt pans can hide traps. There are few things more distressing than the sudden sense of uncertainty that your car transmits through your ass as traction starts to fail.

Other illusions are not as cruel. If you can keep your glasses from misting up, a warm foggy morning can do for you what a wind does for the deaf; it cancels your disadvantage. Birds are often invisible until they are within range. There is that brief moment of galvanic awareness that often as not makes you freeze up as the bird flashes into view and vanishes again. (Did he see me? Is he coming back? If so, from what direction?)

But fog can confuse even the keenest eye. There was the still, misty dawn when Dick Dinkelspiel and Stan spotted through their blind's cover a snow goose sitting at the edge of their pond on Wheeler Island. In breathless anticipation they watched as the bird drifted toward them, until it was near enough that they realized it was smaller than a goose. Was it a mallard? No. A teal? No. Another few minutes and the "goose" turned into a single downy feather, randomly propelled by tiny air currents.

On rare occasions when the sun starts to break through, you can see "fogbows," mysterious glistening arcs of reflected sunlight but pure white, with none of the colors of a rainbow. These are known technically as Ulloa's Rings, and conditions have to be just right for them to appear.

There are other optical illusions that don't depend on fog. At dawn, the hills that rise beyond Route 680 on the west undergo fascinating changes of color, almost seeming to glow with their own internal light against the darker western sky. During the day those same hills always look as if their crests are lying under a cloud's shadow, whether or not there is a cloud to cast it. It's not as dark as Diablo, but some subtle change in soil or rainfall must give the grass its somber color, for all the world as if something had shut out the sun.

The human eye is used to perceiving things from several feet up. In a duck blind, with your eye at water level, the little wavelets of a five-mile-an-hour wind look as if they are roaring past at sixty. They slow with each inch above the water that you lift your head. (And yes, if you try to prove this by experimenting, it will be just when your head is fully exposed that a big drake mallard, unseen by you until the last moment, will flare away with a derisive quack.)

The trains, too, project their own illusion. The roadbed is about as flat as engineering science can make it, but from a few hundred yards off to the side of the tracks, the trains appear to rise up from under the marsh, describe an arc, and then descend back into it. It is as if the curvature of the earth had been shortened to the size of a Mars moon.

Then there is that infuriating little spider idly spinning her invisible web against the sky. She always chooses the gap in the cover that you use to spot approaching birds. Just a moment ago, your distance-focused eyes had you thinking she was two sprig coming in with set wings, and now she has done it again. And if you look away, in three minutes she'll fool you once more. And just when you are primed to ignore her, a *real* pair of sprig will take her place.

To some degree science can explain how normal perceptions can go off the tracks, but in a place as mysterious as the marsh there are others that shade into hallucination and then into the supernatural.

The predawn placement of decoys is a case in point. You leave gun and dog at the blind and wade out into the blackness until they disappear, far behind. When you throw out the dekes, you worry you have gone too far. But at sunup you discover your spread clustered about ten yards out, staring accusingly at you with their blind glass eyes. Were you really that close, or did they quietly swim in behind you?

Another hallucination is the passage of time, whose regularity science may have determined down to the micro-millisecond, but which varies its stride wildly on the marsh. Ask any boy (and a good many grown men, for that matter) how long the night before a shooting day is, and you will get some feel for eternity. Especially for the young, the night simply will not end, no matter what is done to push it along. The odd thing is that you will swear you never even blinked, yet when the alarm goes off it finds you unaccountably asleep. How did that happen?

Then comes breakfast and an abrupt speed-up in the pace. Early in the season, before things get down to a routine, there is a frantic scramble to assemble kits as the first light shows in the east. There follows the race with the dawn—on foot, by car, by boat—to reach your blind and set decoys before shooting time. All preliminaries at last accomplished, you settle in your barrel and check your watch. Six minutes to legal shooting time? Just about right.

And at this point Time, having whipped you along unmercifully for the previous two hours, takes another break. Two mallards in easy range loom out of the dimness and vanish. A pause, followed by a faint splash as some other duck settles in close by. A small company of teal zips past on hissing wings, and a mudhen lumbers over less than a dozen feet up. Off to the left, someone is doing an original form of duck calling that is rattling ducks into the air, while on the right some no-goodniks in the neighboring club are cutting loose with premature barrages. A mosquito whines in your ear and you slap violently.

Time yet? You look, and it is still six minutes to go. Can't be, but it is. In the last moments before legal shooting you could read *War and Peace* twice and still have some seconds left over. But it finally does arrive, and with it a growing volume of gunfire from other blinds. Your first legal chance swings past, banks in as you quack softly, then flares as you rise. The day has begun.

Oddly enough, that does not mean that time speeds up markedly. I am normally pretty accurate in guessing the hour and even the minute but am constantly amazed to find that my estimates are way out in front of the watch hands as the sun angles up the southeastern sky. Later in the morning the flow becomes normal, but those first hours are blessedly slow-moving.

At the end of the day, with sundown relentlessly closing on you, the minutes pass erratically, faster or slower than normal depending in part on how many birds are working into the marsh. Actual official sundown, of course, comes some time after the sun itself has disappeared, but you rarely have trouble knowing when that is. There is something unmistakable about the last, sudden, accelerated dimming. Besides, why else would that mallard choose just this moment to be drifting down the sunset to your decoys?

Hallucination is one thing and the supernatural another. We have a haunted electric clock in our club's kitchen. It is a large, wall-mounted clock with a phony pendulum, and it advertises Schweppes tonic. Many many years ago, long before I became a member, it stopped at 25 minutes after three. Given the corrosive nature of Suisun air, its innards were presumed to be nothing but a mass of rust and green corruption. Yet in December of 1993, the damn thing began to run again, all on its own.

The only trouble is, it's on East Coast time, precisely three hours ahead of us. Who, how, why...? I don't know. Our members all deny responsibility. But I'll tell you this: I won't reset the hands. I won't touch the thing.

The supernatural creation and deletion of ducks in and from the sky is a simple process known to all hunters. This section is for those who have not had the experience.

To create unwanted ducks, it is only necessary to leave the blind. You may stare at the empty sky for a half hour, resisting the ever stronger demands of a filling bladder, or you may even make two or three false starts, in the hopes of decoying birds in. But it is only when you are standing outside the blind, thoroughly committed, that the birds suddenly appear from some other space-time continuum. They pretend to be startled, of course, but it is all an act; they have been patiently waiting in another dimension, perhaps on the planet Zog, ever since that second cup of coffee began to make itself felt. By the time you grab for your shotgun, they are long gone.

Deleting ducks when you want them is no more desirable than creating them when you don't, and it is just as easy and inevitable. A group of birds approaches, wings set but still out of range. Tracking them carefully, you tip your head back slowly as far as it will go, trying to expose as little of your face as possible as they pass overhead. Eventually your neck rebels, and you must turn slowly to keep them in sight. For just an instant, your hat brim obscures them. And then they aren't there. You look left, right, up, down, back around behind you—but no, they are gone, utterly and totally gone.

Where? I think I know. Somewhere at the other end of the galaxy a Zoggian, after scanning the sky and finding it empty of ducks, has just gotten out of his blind and unbuttoned his fly.

If by chance you have prevented the ducks' disappearance by keeping your eye on them the whole time, the Invisible Malignant Neighbor steps in. The marsh has been utterly silent for the past half hour, but just as the birds bank toward you his gun roars, and they flare away. Who does it and why? I know not. You never see either him or what he was shooting at.

In mild weather the sense of touch does not contribute much to your perceptions. There's the occasional mosquito to swat or crick in the neck to massage out, but it's easy to fall into a semi-hypnotic trance while watching an empty October sky or the dancing wavelets. In fact in the early part of the season heat can be almost as big a problem as cold becomes later. Sunburn is likelier than frostbite, and a set of waders can turn into a do-it-yourself sauna. That leak that you didn't repair well enough last season is welcome when it opens up again, even if it does mean some chafing around the ankles. But the cold, when it comes, is something else.

Those same waders (What? You still haven't fixed them?) now let in first one and then other trickles of agony. In the first heavy frost some scum ice may form on the ponds, but it is no problem; you just push through it. But a few really cold days in a row and things can get tough.

It is just at this time that cracks hidden in the marsh floor seem mysteriously to open up. Where you waded with well-earned confidence yesterday, today are cavernous gaps whose nether ends are in some outhouse in China. Usually it is only one leg that goes in, but that is enough to cause real problems. Weighted down with clothes and gear, with one knee up around your ear and the other straight down in the bottomless pit, you wonder just what you are going to do next. (One answer: take along a ski pole to probe for the holes ahead of time or help you out when you miss one.)

In 1991 we had a cold snap that put nearly an inch of ice on the marsh—not enough to support a man or even a big dog, but plenty enough to slow you down and wear your shins raw, even through waders. I finally ended up walking backwards and kicking up under the ice to break it; my dog's solution was to move fast enough not to break through.

Even without wind, icy days move the birds around. Rarely are you skunked when the temperature falls below freezing.

Fog in early fall is one thing; in December, no instrument yet devised by man is capable of measuring its chill factor. From personal experience I know of a few relatively tepid rivals: a snow storm on a Connecticut marsh, for example, or tidewater turned to ice on Maryland's Eastern Shore, or the North Sea in October when you're swimming after a downed bird. But they are not in the same league.

If a warm Suisun fog takes away the duck's advantage of sharp eyes, a cold one leaves all but the hardiest hunter in frozen, aching immobility. His ears are weighted with thorns. Toes lose all feeling except pain. His dripping beak turns bright red and drizzles like a leaking faucet. Just as he gets his hands in a position to warm their fronts or backs (it's never possible to do both at once), birds appear suddenly out of the mist, and his stiffened fingers, as supple as lobsters and colored much the same, refuse to obey whatever orders his fuddled brain gives them. Long after the birds have vanished he is still fumbling after his gun or call. Even if they return and the hunter is blessed with ears that can still hear their wings in time, the fog can send him misleading signals about where the noise is coming from. Usually it is 180 degrees away from where he is looking.

And I wouldn't trade a day of icy Suisun fog for all the sunshine in Hawaii.

Messages from your nose when the weather is warm can be

almost as compelling as those from your skin when it is cold. One could be blind, stone deaf, and numb as a leper, and still appreciate the marsh, just from her smells. Some of these are natural, others man-made, and for non-hunters most fall into the stink category.

My mother used to complain about the smell of Hoppe's No. 9 powder solvent ("Banana oil!" she'd snort disapprovingly when we cleaned our guns at home), and if she had ever gone to the marsh (she never did), she probably would not have liked the smells of outboard exhaust, burnt gunpowder, and old rubber boots either. In Big Stan's club these would combine magically with Angostura bitters, old leather, and many yesterdays' soups. It has been more than fifty years since I smelled that peculiar combination, yet just the memory brings on a wave of nostalgia.

If just a few of the man-made smells are open to criticism, virtually all of the natural ones are socially unacceptable. (The duller a sense perception, the more intolerant it is of extremes. Bright lights offend the nearsighted, loud noises the deaf, and pungent smells the nasally dull, whereas well-developed sense organs seem to revel in extremes. If you don't believe me, interview your dog the next time he has rolled in something long dead.)

The marsh's perfumes are powerful enough that even some of the otherwise delicate odors that blend with them, such as the fall blossoms of coyote brush, become somehow suspect. But it is just that overpowering combination of stinks that sets Suisun off from other waterfowl environments in California. The rice country and grasslands of the Central Valley have their own audio-visual magic, especially the mammoth flights of geese, but they are a sort of flat picture postcard in the absence of the nasal dimension.

A lovely aroma? Well, no. Lots of things died to give the marsh her own peculiar fragrance, and many of them didn't smell too good to start with. (My own personal superstition is that if my first fart of a hunting morning smells like the marsh, it will be a good day; if it fails to meet marsh standards, hunting will be poor.) In any case, you need the warm, still blackness of a dewy October pre-dawn, when there is nothing else to distract you, to get the full impact. After the sun comes up you see strange gases, probably mostly methane, rising from the depths, oily bubbles that pop and leave iridescent rings on the surface. An ebb tide reveals mudflats in the sloughs that are as ripe as they are bottomless.

In my next incarnation I want to be a some hunter's retriever so I can really appreciate all the marsh has to offer a keen nose. My owner will probably beat me for what I roll in.

Anyway, that's why we love her. You don't understand? Go see for yourself.

Appendix B

Wildlife Notes

A detailed study of everything that lives in the Suisun Marsh would be a monster. This section is just about that small slice of bugs, beasts, and birds that have made themselves most felt, however pleasantly or nastily, when I shared their living space.

Bugs

Most people think of duck hunting as something that demands weather so utterly foul that only hunters and dogs (and ducks, of course) are crazy enough to be outside in it. They assume that all other life quietly retires to its version of a good book and an open fire. That's wrong on two counts: a warm fall day does not rule out a good hunt; and, regardless of temperature, Nature may slow but she never stops. No matter what the weather, the Suisun marsh is a great place to see her in action. You even get a chance to participate, not only as predator but as prey.

On hot days early in the season, you slap mosquitoes and check the insides of blinds carefully for black widow spiders, under no circumstances reaching into out-of-the-way places without first examining who might have gotten there first. Although the black widow almost never has enough venom to kill you, the bite is so painful that some people must be hospitalized for that reason alone.

It is especially important to check outdoor toilets for the 'widows, because the most common location of a bite, I am told, is the scrotum. (Now that *would* catch your attention, wouldn't it?)

On the marsh itself, throughout the season there are also huge yellow-and-black-striped garden spiders, at least an inch across in the body alone, that spin webs in the alkali bulrush and other undergrowth, defiantly blocking your path. They look mean, but I have never known anyone to be bitten by one; in any case, they are totally harmless. But the webs look strong as monofilament, and the weight of their sinister builders will bend a stout bulrush. I know grown men who give the horrid beasts a wide berth just in case.

If the spiders flourish during the warm weather, so do their natural enemies, the wasps. They infest most clubhouses and will sometimes produce an out-of-season hatch when the weather turns balmy or someone leaves a heater on too long. Like the spiders, the wasps are more bluff than real threat to humans, buzzing and darting in front of your face but rarely stinging. Oddly enough, they are most dangerous when they are dopey from the cold or an underdose of insecticide and find it less work to sting than to get out of the way.

Be glad, however, that you are not a spider when confronted with a wasp. The wasp strikes with surgical accuracy at the one spot in a spider's anatomy where the sting produces instant paralysis. If the wasp for any reason misses, she (like the spiders, it's the female to look out for) may well end up on the spider's plate for dinner. But at least she will die instantly, whereas if *she* wins the fight—and she usually does—the spider will be put into suspended animation and installed in the wasp's nursery, where it will become a leisurely first breakfast for the wasp's grubs when they hatch. When it comes to wasps and spiders, Nature doesn't play Bambi.

A sure sign that the season is changing comes later in October, when young parachute spiders come floating in to carpet the marsh with their discarded webs. These strands are actually more like a cross between a balloon and a sail than a parachute; they lift their tiny pilots by means of microscopic thermals, then catch the faintest of breezes to coast miles across the marsh and countryside. When the spiders finally come to rest, they cut free from their carriers and go to work building real webs, but the discards give a gleaming, metallic sheen to everything. A short drive there during the flying season provides your car antenna with pennants that stream out behind. Sometimes you see the next hatch climbing to the top of the highest part of your blind cover, testing the wind and temperature for a launch of their own.

Another warm-weather bug is one whose scientific name I do not know. At Tip End, Hal Petrich christened it the Love Bug. It is a peculiarly apt name, because until the weather turns cold, the damn things are everywhere and have an obsession with human skin. Even though they don't bite right away, their hard outer shell, like that of a horsefly, makes you assume instantly that they do, producing an irresistible impulse to swat. The blow rarely kills, however, and the miserable things continue to cling to you with the injured air of a rejected lover. And just when you decide that they really are harmless and can be ignored, one sharpens up his drill and zaps you for a slow-healing bite.

The Love Bug—an apt name indeed. I would like to find a final solution to Love Bugs.

In the winter months, after most other insects have died or gone into slow motion, you find yourself besieged by ladybugs. They are harmless and brightly colored, but in an environment that has taught you to be suspicious of any tickling sensation on your skin, they are a distraction. Last year, one sneaked onto my face mask at the bridge of my nose just after I'd killed three comatose black widows; you wouldn't think any ladybug, even looked at too-close and cross-eyed, could pass for something venomous, but this one managed it. Did that mask fly!

How they stay active and keep their sense of humor after all other insect life has slowed or stopped is a mystery.

And then there is the lowly mosquito in her millions. Throughout Suisun duck hunting history, she has made her presence felt. Early repellants depended more than anything else on the toughness of the hunter. In 1908, serious problems arose for those "who had failed to anoint their hands and faces with the various mixtures supposed to make the human epidermis proof against raids of the longbills—those particular unguents generally are efficacious only when the hunter's olfactory fortitude is stronger than that of the insect, and the latter is no slouch in playing the odor game." In short, most repellants worked better on hunters than mosquitoes. (*B & S*, 31 Oct 1908)

Bernice Huber says that in the 1920s her father used to wear three layers of shirt when working the Teal Club fields on even the hottest days, but the skeeters were of a size and determination that they drilled right through. One opening day in the 1980s at Tip End my brother Stan killed 210 before we fired the first shot, and that was after anointing himself with modern, gentler repellant. ("It's just *sauce Béarnaise* to them," grumbles my father-in-law.)

Until the first frost does in the worst customers (and don't forget that any warm day in winter can still produce a late hatch), they are a plague, especially just before sunup. You hear all sorts of stories about Suisun mosquitoes, and most of them are questionable, but my brother Pete got a specimen last season that CWA had banded by mistake. Must have gotten mixed up in a batch of mallard ducklings.

Mammals

Our most common mammals are jackrabbits, skunks, ground squirrels, mice, and muskrats. When we flood the marsh in the fall, the first four find their range drastically reduced, and they are confined to the remaining bits of dry land, mostly levees and clubhouse premises.

Jack rabbits often strain man-dog relations by tempting our

Best Friend into wild chases, but skunks move relations into the man-dog-*wife* dimension. By itself, a skunk may alter one's appreciation of nature pretty drastically; in connection with a dog, it may threaten tranquility on the home front. There are many ways of removing a skunk's fragrance from a dog, and none of them works, at least not immediately. One is supposed to be Massingill's Douche, but the most tried and true method is tomato juice. I find that unsatisfactory if for no other reason than it makes a Bloody Mary temporarily distasteful—the skunk association somehow remains even with unsullied juice. Besides, have you ever tried to clean up yourself and the premises after a dog has shaken off his skunked tomato juice?

My own favorite is simple vinegar, applied in a strong solution. It works just as well as the red stuff, is cheaper, and it doesn't leave a mess. As for commercial brands I've tried, I find their odor offensive before use and intolerable after mixing with skunk.

One year I gave my annual raffle-winner visitor from England a really horrible initiation into California hunting. On the first day of his visit we got lost in a dense fog on Grizzly Island's public hunting area, his new waders split at the crotch, letting water into both legs, and we never fired a shot. Then I heard an odd bark from my yellow lab Carry, whom I'd put on high ground near by, and that unmistakable aroma wafted over us. Well and truly skunked we were, both figuratively and literally. Moreover, we had a couple of hundred miles' drive ahead of us, and my dilemma was whether to put Carry in my truck's cab with us, where we could wrap her in a disposable blanket, or in the camper shell, where she could share her perfume with our bedding, spare clothes, and food. But the mile walk back to the car, with Carry rolling energetically in the flooded pickleweed and other grasses all the way, somehow did the trick. The skunk was reduced to a fading memory, most of it concentrated in her collar. Maybe Suisun water is the ultimate cleanser.

Muskrats are another pest. Their burrowing through levees makes them instant allies of the tide in its unending war against the land, subverting the best efforts of man and backhoe; a small emergency slough-side exit from the den can be transformed in the course of one high water into a swiftly widening breach that calls for expensive measures.

A muskrat's untimely death at the bottom of your blind—a week-old corpse is just about optimum—provides a new dimension of marsh stink and a nauseating disposal experience. Worst of all, however, is when the beast survives in your barrel long enough to greet you in the predawn blackness. It is an excellent practice to bring along a flashlight so that you can at least identify any such problem before you climb in, but identification is only the beginning. What do you do next?

This question is perhaps best addressed by grasping what *not* to do. First of all, reject the impulse to shoot the rascal. (No, really; I know of several cases.) The costs include the loss of the barrel not just for the day but for the balance of the season, the $750-bill for installing a new one next summer, and (most galling) the well-deserved hee-haws you will endure until you die or move to Siberia.

Then there is the humane approach. I'll never forget offering a muskrat a way out by lowering an empty decoy sack into the barrel in the hopes he would grab it and allow me to fling him free. Unfortunately, he grasped neither the concept nor the motive, but he grasped me, all right, sinking his incisors into my wrist right next to the artery. No more Mr. Nice Guy at that point, and I did him in with a handy oar.

And if you don't have an oar? Well, I'm not sure, but we have some heavy stools in our blinds that might serve for executions. Otherwise I'd be tempted to see whether coexistence worked (an uncle once shared a blind with a quiet fellow for the best part of a morning before he realized he had company)—and resort to stomping if it didn't. Anyway, you might think about making contingency plans.

In a time when more and more species are dying out, it is encouraging to find some on the rebound, particularly the big mammals. By the middle of the 19th century, the local California Grizzly bears that gave Suisun's biggest island its name had been wiped out, and in 1860 Suisun's last native tule elk was shot. But in 1972 a successful transplant of tule elk back to Grizzly Island from the Owens Valley in Southern California has resulted in a roving herd that now, in the absence of bears, must be culled periodically by man to keep it from overrunning the island. They are impressive animals, the biggest running to half a ton.

In the early years of the elk reseeding program, a visiting hunter from the East Coast was given a considerable jolt when, in response to an odd squelching sound, he peered up over the rim of his blind. Moving directly at him and only a few yards off was a gigantic elk with a decoy caught in its antlers. There was a few seconds' stare-down while each side pondered the correct protocol for what was clearly a unique situation for both, and then the elk resumed feeding while making a short, polite detour around the blind. The hunter just exhaled.

The California Grizzly, alas, is well and truly extinct, but in 1991 Bill Frost looked out his window on the island side of the Montezuma Slough bridge, and to his astonishment saw a black bear. A call to Fish and Game produced several wardens armed with a tranquilizer gun, but the unfortunate bear took the dart in a major blood vessel and died before they could treat him. Where did he come from? No one knows, but it was probably from Mt Diablo, where the old grizzlies used to hang out. Yet to get to the island from there or any other known bear habitat these days, it is necessary not only to swim the Suisun Bay or the Montezuma Slough but to cross at least one major highway beforehand. (How would you like to find *that* suddenly in your headlights in the fast lane?)

At the other end of the mammal spectrum is the endangered Harvest Mouse. I have never seen a harvest mouse, at least not one that I recognized as such, nor do I feel necessarily deprived of a great experience. They doubtless feel the same about me. Moreover, I am certain they would not take as many pains to ensure mankind's survival as mankind is taking to ensure theirs. Nevertheless, I applaud the effort to keep them going. The only thing is, I wish that the mouse could manage more on its own, without the aid of the horde of well-meaning but largely marsh-ignorant bureaucrats, who, in the mouse's name, demand a say in every shovelful of earth turned there. No fault of yours, mouse, but the human company you keep has some drawbacks, at least among us two-legged marsh rats.

Ducks and Mudhens

Bird life—at least duck life—on the marsh is what prompted this book, but ducks are only part of the avian population. Moreover, so much has been written about snob-appeal waterfowl (sprig, mallard, teal) that it is only fair to start with the humble Untouchables at the bottom of the duck social scale: spoonies and mudhens.

The lovely, lowly spoony, or shoveler, has the reputation of being unqualified as a target, unworthy of a hunter's guile, and unfit for human consumption. One of the earliest watersheds of hunting snobbism is the decision to avoid shooting any more spoonies. This is not always an easy vow to observe, for, like the gadwall, the spoony often masquerades successfully as a mallard until the last possible instant before the trigger is pulled. Moreover, as my late uncle Roger Kent used to say, spoonies have a death wish; shoot anywhere in the same quadrant of the sky and they will fold.

Imagine a beautifully costumed, exceptionally talented ballerina with a lovely figure—but with a nose-dominated face that would stop a clock, a bit of an alcohol problem, and a bad case of BO. That's a spoony. From the forehead aft, there are few ducks to match its brilliant plumage and clean lines, but there is something about that spatulate bill that turns the whole bird into a bad joke. Though sometimes capable of beautifully dramatic aerobatics, it often seems to fly just a little bit out of control, like a serious drunk adjusting to a teetering world. And the gourmet qualities of a spoony that has fed on the wrong things—and they often do—would discourage a hungry seagull.

Oddly enough, there were hunters (or at least writers) in the early days who rated the spoony very highly. John X. De Witt, hunting editor of *The Breeder and Sportsman*, was one of those who often referred longingly to the spoony's succulence, and the old market hunter Walter R. Welch wrote that in the 1870s and 1880s, he used to get the same for spoonies that he did for widgeon ($1.25 to $1.50 per dozen) and more than for some teal ($1.00 to $1.75 a dozen). (Welch, 1927, p. 183)

The only Suisun waterfowl lower on the social scale than a spoony is a mudhen. This is the nickname that has stuck most firmly to the coot, but it is not the only one. He has been saddled with many others, ranging from the sarcastic (black mallard, black teal, Petaluma mallard) to the downright defamatory (crow duck).

From the late 1890s, Edward Dinkelspiel used mudhens in the *Solano Republican* to poke fun at his tyro hunter friends: "S. M. Getchell has been on the marsh twice this week & returned both times with long strings of black mallards & other choice ducks." (*SR,* 15 Jan 1897) "Dr. Bailey & Leonard Prior made a third visit to the famous Rosenberg ponds Tuesday and again succeeded in killing four of the variety of birds which abounded there. They are inbred black mallards and at this season of the year are fat and juicy." (*SR,* 15 Jan 1897; 18 Jan 1901) (But it is interesting that all three of these gentlemen later turned into discriminating duck hunters.) *The Breeder and Sportsman* wrote about E. Klevesahl, also a future excellent shot, having "covered himself with glory and coots"—so many out of one flock, in fact, that "his retrievers took fright and started for home." (*B & S,* 21 Oct 1899)

The mudhen was not without his defenders, though the motives of some were suspect.

An impassioned defence of "the great American coot, miscalled a mudhen" as a "game bird and table delicacy" appeared in a 1905 edition of *The Breeder and Sportsman.* With great earnestness, the unnamed author (who goes on using the term mudhen, incidentally) notes that the bird does not feed on mud, does not like mud, and, given the choice, will not live in mud. Well, yes, if there is no other food available, it will feed on fish, but so will any other duck. Mostly, however, it feeds on wild rice and has a "delicate and delicious flavor." In fact, a fancy San Diego restaurant puts the mudhen at the top of its menu, referring to it as "poule dou—water hens" (sic—the unnamed restaurant was as good linguistically as it was gastronomically). Not only that, but the mudhen is a challenging target: "Speeding down the wind, it is at least as hard to center as... sawbills [mergansers], for example, or spoonbills, or gadwalls." (Ah, now we know the class of hunter the author is aiming at.) And finally the truth comes out: "Mudhens are the bête noire of the club grounds, for they have a capacity for, and persistent cunning in, cleaning out the baited ponds of the food intended for the ducks." In fact, they are "becoming a nuisance," but when everyone realizes what a great game bird they are, they "will become comparatively scarce," just like other ducks. (So hurry.) (*B & S,* 30 Dec 1905)[1]

Somewhat less subtly, the Fish and Wildlife Service in recent decades tried to rename mudhens the Whitebill (one of the few compliments the poor bird has ever received), but it never caught on. Again, the hope was that a different label would turn the creature into a noble quarry that hunters would pursue, thus ridding marshes of a food-gobbling pest. But the original humble name stuck, and with it a measure of security for its bearer.

In 1908 the Family Club invited one Pete Sloan, a novice

---

[1] Rod Williams, an old hunting crony, tells of seeing lazy mudhens let widgeon do the foraging for them. The widgeon, having dipped down and torn food from the pond bottom, would tip upright, only to have their meals torn away by the waiting white-billed thieves.

hunter, for a shoot. Friday evening, all the talk was of Black Teal and their superb qualities. Pete took it all in, and next dawn found himself in a mudhen paradise. "On the wing, on the water, perching on the tules, any old place, he got them—twenty-seven of them," and he could have had more but had been warned not to shoot over the limit. The charade continued on his return to the clubhouse, and he bore his bag back to the city, where he held a gourmet Black Teal lunch. The Family Club, however, was taken to task for having thus nipped at the bud the "latent hunting ambition of one who might some day want to shoot elephants or craps." (*B & S,* 17 Oct 1908)

I must confess my own fondness for mudhens. There is something heroic about the way they flail their way aloft in a mad sprint, leaving footprints of suspended water in their wake so fast that the highest ones are about six splashes aft of their retreating tails. They finally get airborne on sheer willpower, clawing their way up to a ceiling of about twenty feet before they level off to lumber onward for perhaps 200 yards before sprawling exhausted back onto the water. Yet each spring and fall they somehow manage to migrate for thousands of miles, keeping their real endurance secret by flying at night.

One of the funniest sights on the marsh is a mudhen landing on ice, where the normal sudden-stop splashdown becomes a long and hilariously uncontrolled slide. I never used to think the bird had any self-respect whatever, but the look of wounded dignity when one finally comes to a stop and rights himself has convinced me they have some pride, however misplaced.

When I was a child, we had a Chinese gardener named Sling. We would deposit on Sling's doorstep whatever owls, crows, hawks, or other victims of our indiscriminate guns we though he might like. He welcomed all in his pidgin English as "Hip good—make good medicine"—all except the mudhens: "Mod chicken hip no good," he'd say. The ultimate insult.

Probably the most desired duck on the marsh is the pintail, or sprig. Up through the early 1990s a series of disastrous nesting years shrank the sprig population to a shadow of its former grandeur, but in just a few years they have now made a great comeback. In the 1950s there were so many sprig that hunters were allowed up to four "bonus birds" over the limit for other waterfowl, but in 1996 you were allowed only two sprig in a limit of seven ducks overall.

With their long, lean body profile and gracefully thin wings, sprig are aerodynamically cleaner than any other puddle duck, and their habit of working over an entire club's acreage before deciding to set in produces a unique mixure of anticipation, appreciation, and frustration in the waiting gunner.

For sheer aerial ballet, nothing matches the courtship flights that begin each mating year as early as December or even November. One hen, pursued by up to a dozen drakes, will lead them on wild, whistling, three-dimensional chases, banking and diving down to marsh level, then soaring hundreds of yards up. Often, you see one disconsolate drake following morosely some distance behind. (Or is he the smart one, waiting till the show-offs have exhausted themselves?) Every now and then the hen will land to give the swains a 30-second break before she leaps skyward again.

By late January, the sprig are done with highschool romancing and starting to settle down to serious relationships. Given the huge mortality among hens during the drought years (they were easy meals for predators as they sat on exposed nests), there was a heavy imbalance between them and the drakes. Perhaps as a result, on our club there was a neat and consistent grouping of two drakes to each hen as they prepared to fly north a couple of years ago. I hope each two took proper care of their lady.

The next most popular bird—and in first place for many—is the mallard. (My sister-in-law, whose father hunted only mallards in his native Oregon, greeted Stan on his first return from the field with a nice limit of sprig with the words, "Yes, but didn't you get any *real* ducks?")

Studies by the California Waterfowl Association show that more than 80 percent of the mallards shot in California are hatched in California, and recent efforts to encourage local nesting have been paying big dividends in increased mallard numbers. So far, however, efforts to convince federal authorities that California should have a bonus mallard have fallen on deaf ears.

Seen from afar, a mallard on the wing tends to look a bit bottom heavy, with his tail end sagging from all that weight. But he makes a handsome trophy and a full meal.

The acrobat of the marsh is the little teal, who jinks around at low altitude, alone or in small groups or occasionally in swarms, and always manages to come at you from unexpected angles. He is the smallest and most active of our puddle ducks, darting around the marsh long after the rising sun has led other birds to find a quiet resting spot.

Our two most common varieties of teal are the greenwing and the cinnamon, with the latter much rarer during the hunting season than the former. Most cinnamons arrive before the season begins here and return after it closes. Blue wing teal are also occasional visitors in the off season.

A century ago widgeon and canvasback easily outnumbered all the other ducks, and widgeon are still among the most common on some clubs. They are often maligned as a lesser bird for the table, but I have found very few to criticize on that score. Most of ours have the white fat associated with clean living and a proper diet. Their piercing whistle is a standard part of the night symphony on the marsh. Like the sprig, they have noisy courtship flights during the late season.

In recent years the gadwall has become more and more common. In poor visibility it is easily confused with a mallard, which is unfortunate for the gadwall. (Most gunners, given the choice, would not voluntarily shoot the bird, which often tastes of mud. But there's no profit for the gadwall in tasting bad once the mistake has been made.)

Late in the season, in stormy weather and on deep ponds, the bay ducks sometimes come into the marsh as refugees from the outer waters: bluebill (scaup), redheads, and canvasbacks. The latter two, being vegetarians, make good eating, but the bluebill, who usually lives on more active marine life, if cooked conventionally should usually be served with a good white wine and a clothespin for the nose.

We also get resident flocks of bufflehead and ruddy ducks, which, because of their diet, almost no one shoots. The ruddies, somewhat arrogant in their sense of security, fly in large flocks right down on the deck. They delight in buzzing you on high-speed strafing runs so low that one of our club members has tried—unsuccessfully so far—to club one down with his gun barrel. (I'm glad he's missed—they're an endearing, feisty little bird, and their hedgehopping at full bore is a delight to observe.)

Geese and Other Waterfowl

Not many Suisun clubs get good goose shooting nowadays. We have the standard California species: Canada geese in all their variations, from cackler on up; snows and Ross' geese; and whitefronts. The Canadas are starting to pick up in numbers as ever more of them stay around to nest locally instead of flying north in the summer. The others come visiting from the Central Valley, but in nowhere near the numbers seen in the valley itself.

At the edge of the waterfowl definition (they don't swim, after all) is the jacksnipe, a bird that in the old days was a favorite game bird when the ducks weren't flying. We still get some coming through on migration, but in nowhere near the numbers seen in the 1890s. With a very fast, corkscrewing takeoff and a camouflage pattern that almost defies detection, you sometimes can only hear its rasping peep and perhaps get a glimpse of a white feather against the marsh background.

Pheasants are happy in the marsh's habitat, though most of them are probably spillovers from commercial pheasant farms or clubs that have planted them. I was once told that they fed on some wild plants that gave them a bad flavor, but I suspect my informant was hoping just to discourage me from reaping part of the harvest. So far, I haven't had a bad one.

Off the game bird list come many others, including our biggest waterfowl, the swan. There is a majesty to a flight of these great white birds, especially when you see them highlighted against dark gray clouds, and their plaintive, never-answered question (Who? they ask, Who?) haunts you long after they've disappeared over the horizon. Perhaps they are asking the identity of the last jackass to shoot at them, an event that has probably happened a few times that day already.

Fortunately for the swan, however, most hunters who make that mistake have no idea of range, and the birds suffer relatively few casualties from gunfire. The birds are making such a comeback these days that some states now have a limited open season on them, but there is little excuse to hunt them unless you are starving and are equipped with Neanderthal jaws; a swan is mostly rawhide.

I guess he has to be. Whereas herons and pelicans solve their out-front weight problem by bending their necks into an S-shape, the airborne swan makes no such compromise; *his* neck runs about a yard straight out in front of him, with his head perched at the end. Now granted that swans are known more for bad temper than braininess, but their heads have to weigh *something*, and how in the world can they fly hundreds of miles nonstop that way? Just try holding a big lemon (about the size of a swan's head) at arm's length in front of you for ten minutes to see what I mean.

Surely one of the least beautiful things to wear feathers, the cormorant's only social value is to other cormorants. Unfortunately they are good enough at it that are far too many of them. Aside from that, their only function is to hover at the edge of your vision, pretending to be a duck, and to deplete whatever fish stocks are in your hunting area. It is, of course, illegal to shoot a cormorant (why—are they considered a songbird?), but that is a mistake. Somehow, they should be culled periodically.

For the gunner who is pledged to eat whatever he shoots—and that should be all of us—there is no such thing as an inedible bird (yes, even mudhens). Granted, I have not tried a cormorant, but the closest legal thing to that unlovely creature is a surf scoter, and we have found no problem in serving them to discriminating palates. The basic secret in preparing any bird accused of unsavory eating habits is to remove the skin and every other trace of fat. Once free of fat, the meat can do with a red-wine marinade, but not even that is mandatory. The bird may never equal a mallard, but he still beats anything served in a fast-food chain.

Herons, Raptors, and All Others.

As boys, we Arnolds divided birds into two main categories: prey, which we honored by recognizing their species, and almost all others, which, regardless of other attributes, we lumped as Pee-birds. The first category, I regret to say, included all hawks and most owls, damned as competitors in the hunting business. Jays and crows also fell into this group, but I have no pangs of conscience for having shot them; they have multiplied disproportionately thanks to man, and man should cut back on their numbers. Same goes for cowbirds.

But with age and experience, I no longer try to set Nature straight, even with crows. For me, things with feathers are mostly just for watching (even if I don't always care about their labels), and occasionally for shooting (where I always do).

On the marsh, the non-game bird population is huge and fascinating. It ranges in size from hummingbirds to the pelicans and eagles that now and then use our thermals. During the day, the eucalyptus trees around the clubhouse provide particularly attractive habitat to hummingbirds, red-tailed hawks, and blackbirds. As darkness falls, the gumtrees become home to the snowy egrets and great egrets that coast ghostily in to roost just as the last of the sunset fades. Then it's the turn of the owls, with a great horned offering what sounds like advice, and a barn owl who screams his outrage whenever you walk near his tree and whose wing beat is the thing of nightmares as he leaves for a hunt. (I can think of nothing more terrifying than being a mouse *that* fellow has chosen for dinner.)

The daytime raptors include sparrow hawks, sharpshins, an occasional peregrine falcon, kites, harriers, rough-legged and redtail hawks, and even an occasional eagle. The harriers, or marsh hawks, are not only beautiful but practical to watch. Otherwise irretrievable wounded birds can often be located by using the marsh hawk's eyes: when you see him land close to where your duck disappeared, mark the spot and get over there. (But you'd better hurry, or all you will find is feathers.)

Songbirds and waders, including snipe, like the tiny islands around your barrel blinds, and you can get some wonderful closeup views of them, especially on foggy days when they seem fearless. Whether or not it's foggy, a common visitor is the marsh wren, whose survival techniques include building a number of false nests in the tules and cattails to mislead predators. She has a wonderful disdain for up and down. Her position depends only on the orientation of her perch: sideways for a vertical reed, either right-side-up or upside-down for a horizontal one, and everything in between. She will come within inches of your nose, and I've even had one light momentarily on my sleeve.

But some little birds are downright insulting. There is nothing more calculated to make even the most realistic decoy look foolish than to have a small bird use its back as a dinner table, as a pulpit for claiming territory, and finally as a bathroom.

In the late fall, looking far off across the marsh you can sometimes see what looks like a black haze that changes shape, disappears, and reforms as you watch. It usually turns out to be migrating blackbirds in the tens of thousands. The line of flight can stretch for a mile or more, and when they pass overhead, they can literally darken the sky.

But whatever other bird life a hunter may see during the day, what parades across his closed eyes that night are the ducks.

✧

APPENDIX C—Seasons and Limits: 1852-1996

(NOTE: All restrictions carry over to subsequent years unless specifically changed)

Year	Open	Close	Days	Daily Limit	Pos-session	Comment
1852	20 Sep	1 Mar	162	None	None	Season only applies to Mallard and Wood Duck in certain counties. **Note: Only the closed season was defined until some time in the 1900s.**
1853	1 Sep	20 Mar	201	None	None	
1854-1860	15 Sep	1 Mar	166	None	None	(One bonus day on Leap Years.)
1861-1877	15 Sep	15 Mar	181	None	None	Season applies to Mallard, Wood Duck, Teal, Spoonbill, and all other "broadbill" ducks.
1878-1879	15 Sep	15 Mar	181	None	None	Gadwall, Redhead, Cinnamon Teal added to list.
1880-1882	15 Sep	16 Mar	182	None	None	Season applies to all duck species in all counties.

Year	Open	Close	Days	Limit	Posses.	Comment
1883-1887	--	--	365	None	None	No season or protection. (Ducks not listed among game birds that have closed seasons.)
1888-1890	15 Sep	15 Mar	182	None	None	Cordelia Club by-laws forbid Saturday shooting, shooting more than 2 consecutive days in a week by any one member, and shooting after sundown.
1891-1892	1 Oct	1 Mar	152	None	None	San Francisco County prohibits hunting, possession, or sale of ducks or snipe 1 Mar to 1 Sep.
1893	1 Sep	1 Mar	182	None	None	Earliest opener since 1853. Never repeated. State law against using guns larger than 10 ga. Non-resident lessees on Suisun Marsh set voluntary 1 Oct opener.
1894	15 Oct	15 Feb	123	None	None	Season opens at 00:01 on 15 Oct. Sales of game permitted only during open season.
1895-1896	15 Oct	15 Feb	123	None	None	Sale of game (including ducks) permitted only between 15 Nov and 15 Jan.
1897-1899	1 Oct	1 Mar	151	None	None	Preserves applauded for setting aside only certain days for shooting.

1900	1 Oct	1 Feb	123	None	None	
1901-1902	1 Oct	1 Feb	123	50	--	First official limit. Shooting hours ½ hr before sunrise to ½ hr after sunset. Possession unspecified.
1903	15 Oct	15 Feb	123	50	50	
1904	15 Oct	16 Feb	124	50	50	Some hunters' groups suggest 25-bird limit.
1905-1906	15 Oct	15 Feb	123	50	50	Possession limit also 50. Fines to $500, jail terms to 6 months for game violations. Only waterfowl (not upland birds) may be sold on market.
1907-1908	1 Oct	15 Feb	138	35	35	1907 is also first year of statewide hunting licenses, to be made and issued by each county.
1909	1 Oct	15 Feb	138	25	25	First year of new limit. Shooting from power boats and from behind animals outlawed.
1910	15 Oct	15 Feb	124	25	25	Closed season begins 16 Feb.
1911-1913	15 Oct	1 Mar	138	25	25	Closed season from 2 Mar. Fed law mandates sunrise-to-sunset shooting day in 1913, overruling CA half hours before and after.

Year	Open	Close	Days	Limit	Posses.	Comment
1914	14 Oct	31 Jan	110	25	25	
1915-1916	15 Oct	31 Jan	109	25	25	Wood Ducks protected. State law now conforms to Federal Migratory Bird Law. Weekly (dawn Sunday to dawn Sunday) bag limits of 50 for ducks (precursor of possession limits?).
1917-1920	15 Oct	31 Jan	109	25	(None?)	Sale of ducks prohibited by federal law in 1917. From 1918, shooting allowed from ½ hr before sunrise to sunset.
1921-1929	1 Oct	15 Jan	107	25	(None?)	No hunting allowed from sailboats, aircraft, or automobiles in 1923.
1930	1 Oct	15 Jan	107	15? (25?)	30? (None?)	Federal laws, now in effect under Migratory Bird Treaty Act, set limit at 15. CA law said 25.
1931	16 Nov	15 Dec	30	15	30	Drastic reduction in season length due to drought. Noon season openers inaugurated.
1932	1 Nov	31 Dec	61	15	30	Limit of 15 birds in the aggregate but at least 5 had to be Mallards, Sprig, or Widgeon. Buffleheads and Ruddy Ducks protected. Not more than 25 live decoys. Baiting prohibited.

1933	1 Nov	31 Dec	61	12	24	At least four in a limit had to be Mallards, Sprig, or Widgeon.
1934	20 Oct	30 Dec	22	12	24	Starting time 7:00 AM. At least five in a limit had to be Mallards, Sprig, or Widgeon. Shooting on Sat and Sun only (hence season effectively only 22 days long).
1935	20 Nov	19 Dec	30	10	10	Shooting hours 7:00 AM-4:00 PM. Use of live decoys prohibited.
1936	1 Nov	30 Nov	30	10	10	Protected species: Wood Duck, Ruddy, Bufflehead, Canvasback, Redhead.
1937	27 Nov	26 Dec	30	10	10	
1938	15 Oct	28 Nov	45	10	20	No more than three Canvasbacks, Buffleheads, Redheads, or Ruddies in the aggregate allowed.
1939	22 Oct	5 Dec	44	10	20	
1940	16 Oct	14 Dec	60	10	20	Shooting hours sunrise to 4 PM.
1941	16 Oct	14 Dec	60	10	20	Three-bird limit now applies only to Redheads, Buffleheads.

Year	Open	Close	Days	Limit	Posses	Comment
1942	15 Oct	23 Dec	70	10	20	Shooting hours sunrise to sunset. One Wood Duck allowed. (First since 1915.)
1943	15 Oct	31 Dec	78	10	20	Shooting hours ½ hour before sunrise to sunset.
1944	14 Oct	1 Jan	80	10+5	20+10	Restrictions on Redheads, Buffleheads lifted. "Bonus birds" must be Mallards, Sprig, or Widgeon.
1945	2 Nov	20 Jan	80	10	20	No more bonus birds.
1946	26 Oct	9 Dec	45	7	14	Shooting time ½ hour before sunrise to ½ hour before sunset.
1947	21 Oct 23 Dec	3 Nov 5 Jan	28	4	8	First split season. Noon openers return, then sunrise to one hour before sunset. Optional straight-through season of 30 days permitted.
1948	15 Oct 23 Dec	31 Oct 8 Jan	34	5	10	After openers, ½ hour before sunrise to one hour before sunset. Optional straight-through 40-day season.
1949	21 Oct 19 Dec	9 Nov 5 Jan	40	5	10	Optional straight-through 50-day season.

1950	20 Oct 15 Dec	10 Nov 7 Jan	44	6	6	Optional straight-through 55-day season.
1951	26 Oct	24 Dec	60	6	6	
1952	24 Oct	1 Jan	70	6+2	6+2	Bonus birds must be Sprig or Widgeon, singly or in aggregate.
1953	16 Oct 8 Dec	18 Nov 10 Jan	68	7+4	7+4	Closing time now sunset.
1954	9 Oct 6 Dec	13 Nov 10 Jan	72	7+3	7+3	
1955	15 Oct 10 Dec	18 Nov 15 Jan	72	7+3	7+3	End of noon openers. Shooting times for all days ½ hour before sunrise to sunset.
1956	20 Oct	7 Jan	80	7+3	7+3	
1957	12 Oct	14 Jan	95	6+3	6+3	Longest season since 1930.
1958	11 Oct	13 Jan	95	6+4	6+4	
1959	7 Oct	8 Jan	94	5	10	Noon openers reinstated. After opener, shooting time sunrise to sundown.

Year	Open	Close	Days	Limit	Posses.	Comment
1960	15 Oct 10 Dec	20 Nov 8 Jan	67	6	6	Downward slide. Tradition of Saturday opener established. Closed season on Canvasbacks, Redheads.
1961	14 Oct 9 Dec	20 Nov 7 Jan	68	5	5	Bottoming out.
1962	13 Oct 8 Dec	19 Nov 6 Jan	68	5	5	Wood Duck limit raised to two.
1963	23 Oct	5 Jan	75	6	6	Last noon opener, after which shooting time ½ hour before sunrise to sunset.
1964	24 Oct	6 Jan	75	6	6	Noon openers end. Two Canvasbacks and Redheads allowed.
1965	23 Oct	5 Jan	75	5	10	No more than three each or in the aggregate of Sprig and Mallard. Redheads closed.
1966	22 Oct	4 Jan	75	7	7	Mallard, Sprig, Canvasback restrictions lifted.
1967	14 Oct	7 Jan	86	6	6	Not more than two Canvasbacks.
1968	19 Oct	12 Jan	86	6	6	Not more than three Mallards.

1969	18 Oct	11 Jan	86	6	6	Mallard restriction lifted.
1970	17 Oct	17 Jan	93	7	7	Canvasback restriction lifted.
1971	16 Oct	16 Jan	93	7	7	Not more than two Canvasbacks.
1972	14 Oct 27 Dec	20 Dec 20 Jan	93	7	7	Canvasbacks closed.
1973	20 Oct	20 Jan	93	6	6	One Canvasback allowed.
1974	19 Oct	19 Jan	93	5+2	10+4	"Bonus birds" must be Sprig. Two Redheads allowed.
1975	18 Oct	18 Jan	93	7	14	No more bonus birds. Two Canvasbacks or Redheads in the aggregate.
1976	23 Oct	23 Jan	93	7	14	
1977	22 Oct	22 Jan	93	7	14	
1978	21 Oct	21 Jan	93	7	14	
1979	20 Oct	20 Jan	93	7	14	
1980	18 Oct	18 Jan	93	7	14	
1981	17 Oct	17 Jan	93	7	14	

Year	Open	Close	Days	Limit	Pos- sesion	Comment
1982	23 Oct	23 Jan	93	7	14	
1983	22 Oct	22 Jan	93	7	14	End of the ten-year run of 7-bird limits.
1984	20 Oct	13 Jan	86	5	10	
1985	26 Oct	12 Jan	79	5	10	One hen Mallard and one hen Sprig per day.
1986	25 Oct	11 Jan	79	5	10	Not more than 4 Mallards or 4 Sprig, or 5 in the aggregate. Not more than 1 hen of each.
1987	25 Oct	10 Jan	79	5	10	
1988	22 Oct 4 Dec	13 Nov 8 Jan	59	4	8	Shooting time sunrise to sunset. Not more than 3 Mallard, 1 Sprig. Canvasback closed.
1989	28 Oct 2 Dec	19 Nov 6 Jan	59	4	8	Not more than 3 Mallards (1 hen), 1 Sprig. Two Canvasbacks/Redheads in the aggregate. Shooting time ½ hour before sunrise to sunset.
1990	27 Oct 1 Dec	18 Nov 5 Jan	59	4	8	

1991	26 Oct 30 Nov	16 Nov 5 Jan	59	4	8	
1992	24 Oct 5 Dec	14 Nov 10 Jan	59	4	8	
1993	23 Oct 11 Dec	13 Nov 16 Jan	59	4	8	
1994	22 Oct 3 Dec	5 Nov 15 Jan	59	5	10	Not more than 4 Mallards (1 hen), 1 Sprig.
1995	21 Oct	21 Jan	93	6	12	Drake Mallards open. Not more than 1 hen Mallard, 2 Sprig, 1 Canvasback, 2 Redheads.
1996	19 Oct	19 Jan	93	7	14	
From **1997**	**here**	**on,**	**fill**	**in**	**your**	**own**
1998						
1999						

Annotated Bibliography

(Boldfaced portions of the works cited in this non-scholarly index of sources refer to their abbreviations as they appear in the text. Only sources specifically cited have been listed. There were many more.)

B & S: *The Breeder and (&) Sportsman,* San Francisco weekly published from 1882 until 1918, when it became a monthly.

Boatwright, Dan. *Grizzly Island and the Suisun Marsh: A Critical Assessment.* Nov, 1989 (no publisher).

CA lifornia Legal **Code, 1893**.

California Game "Marked Down". 1896 pamphlet printed by the Southern Pacific RR Co.

Cook, Stuart. "The Unfulfilled Bay." *Waterfront Age* 1, no. 4, Fall 1985.

Coon, William. "The Suisun Marsh -- Threats Endless." ***West Wind,* Oct 1994, Jan, 1995**.

— "A Story of the Past." ***West Wind,* Jul 1994**.

— "Will It Never End?" ***West Wind,* Apr 1996**.

— Interviews, 1994, 1995.

CR: Solano County **Records**, including property transfers, leases, taxes, incorporations, and maps.

Craven, Ike. Speech delivered on 18 Sep 1978 to a meeting of the Sacramento Delta Historical Society.

De Witt, John. "Duck Shooting in California." *Overland Monthly,* 56, no. 5 (Nov 1910): 438-449. De Witt was also the outdoors writer for *The Breeder and Sportsman*, and probably wrote under the pen name "Montezuma" for *Western Field.*

Dinkelspiel papers. Information from family records kindly provided to me by Dick Dinkelspiel.

Division of Water Resources. Bulletins **27** ("Variation and Control of Salinity in the Sacramento and San Joaquin Delta and Upper San Francisco Bay")and **28** ("Economic Aspects of a Salt Water Barrier below Confluence of the Sacramento and San Joaquin Rivers"). Put out in 1931 by the Department of Public Works, these landmark scientific studies showed the harm that could come of bay damming projects and "development" of the delta.

Dysart, Ronald R. *History of Black Mallard Duck Club: Stories and Events of Wheeler Island Land Co.* Self-published, 1990.

Fish & Game Code (Section of California Legal Code). **1893**, **1907**, **1909**, **1914-1915**, **1923-1925**.

Frohrib, Melvin. Interviews with a good friend and a man who has hunted the Suisun Marsh for more than 50 years.

Frost, Joan. *A Brief Pictorial History of Grizzly Island*, self-published in the 1970s and a unique photographic compilation of Grizzly Island and the western marsh.

Gilliam, Harold. *San Francisco Bay,* New York: Doubleday, 1957.

Huber, Bernice. Interviews with a wonderful lady whose memories of the marsh go back to 1923.

Johnson, Patti J. "Patwin" in *Handbook of North American Indians.* 8: pp 350-360, Washington: Smithsonian Institution, 1978.

Keeler, Jack. Interviews with a fine gentleman who has a store of dog stories and useful maps.

McAllister, M. Hall. McAlister's hunting **scrapbook**, with materials from the late 1870s to the 1940s, a fascinating journal.

— "Early History of Duck Clubs in California." *California Fish and Game* 16, no. 4, October **1930**.

— "Recollections of Days of Happiness." *Pacific Sportsman,* Jan **1932**.

— "Suisun Marsh History: The Poachers' War." *Pacific Sportsman,* Mar **1938**.

Moffat, Frances. *Dancing on the Brink of the World: The Rise and Fall of San Francisco Society.* N Y: G. P. Putnam Sons, 1977.

Nadeau, Bart. Interviews with and materials from this most helpful chief archivist at the Western Railroad Museum, Suisun City.

PLF: Pacific **L**egal **F**und. "U. S. Fish and Wildlife Service Denies State Authority over Waterfowl Hunting (news release of 2 June 1987).

Richards, William W. "Place Names of Famous Shooting Ponds." *California Fish and Game* 17, no. 4, Oct **1931**.

— "Fifty Years Ago on the Suisun Marsh." *California Fish and Game* 18 no. 1, Jan **1932**.

SAC: **S**portsmen's **A**ssociation of **C**alifornia, **Constitution and Bylaws, 1891**.

SFC: *San Francisco **C**hronicle.*

*San Francisco **Call**.* Extinct SF daily.

SFX: San Francisco Examiner.

SN: **S**acramento **N**orthern memoranda and letters.

***SR**: Solano **R**epublican.* Weekly paper published in Fairfield.

SSCD.1: Suisun **S**oil **C**onservation **D**istrict. *Final Report, Phase* **1**. Consultant Philip H. Arend wrote most of this.

SSCD.2: Suisun **S**oil **C**onservation **D**istrict. *Final Report, Phase* **2**.

Steel, Marshall, Jr. *A Chronological History of Greenhead, Inc.* Self-published, ca. 1989.

Stoner, Emerson A. "A Record of Twenty-Five Years of Wildfowl Shooting on the Suisun Marsh, California." *The Condor* 39, Nov-Dec **1937**. Invaluable statistics from 1882 to 1907.

Swett, Ira et al. *The Sacramento Northern.* Glendale: Interurban Press, **1981**.

Welch, Walter R. "Hunting and Fishing Reminiscences." *California Fish and Game* 13 no 3, Jul **1927**.

— "Today and Yesteryear." *California Fish and Game* 16 no 3, Jul **1930**.

— "Game Reminiscences of Yesteryear." *California Fish and Game* 17 no 3, July **1931**.

WF: *Western Field.* San Francisco hunting and fishing magazine of the early 20th century.

Wichel, Ernest D. "The Suisun 'Sink'." *Times Herald,* **9 Jan 1966**.

— "City of Newport." *Times Herald,* **26 Jun 1966**.

Supplement: 1996 Duck Club Gazetteer
(See maps on pp. 256-257)

A & B #916
Annie Mason Pt #801
Antioch Golden Eye #420
Arnold Ranch #415
Balboa Farms #525
Bent Barrel #516
Black Dog #609
Blue Bird #903
Blue-Winged Teal #918
Boles Pond #633
Bulrush Farms #530
California Farms #425
Can-Can #526/601
Club 914 #914
Club Lanzo #923
Concord Farms #932
Cordelia #404
Cygnus #418
Dante Farms #910
Delta Farms #931
Delta King Ranch #527
Denverton Land Co. #304/306
Drake Sprig #414
Duck-A-Go-Go #607
Duck & R. #323
Family #423
Fat Hen Farm #124
Fin & Feathers #915
Fleetside Farms #715
Flying D #329
Four Winds #506
Franciscan. Marsh View #408/9
Friendly Godfther #703
Frost Slough #623
Gang Bang #531
Garben Ranch #509
Gibson Horseshoe. #410
Golden Gate #402
Good Year #701
Goodyear Land Devel Co. #707
Gray Island #805/806
Green Lodge #411
Greenhead #321
Grey Goose #122
Grizz/Fizz #507
Grizzly #502
Grizzly Fairview #528
Grizzly Hilton #634
Grizzly King #513
Grizzly Ranch #520
Gum Tree Farm #504
Gunn Ranch #326
Hi Gunner #928
Hidden Cove #618
Hit & Miss #930
Hollywood #132
Honker #538
Honker Bay #902
Honker Bay Farms #912
Honkers #510
Hook n' Trigger #913
Horse Shoe #619/620
Ibis #407
Island #501
Island Duck #907
Island Gun #917
Jacksnipe #131
Joice Island Gun #424
Joice Is. Mallard #220
Lang Tule Ranch #207
Little West Wind #508
Long Point West #515
Machado #616
Mallard Farms #905
Mallard Haven #706
Mallard Inn #117
Marsh #536
Maupin Ranch #627
Meins Landing #631
Mendoza #518
Meridian #612
Miramonte #419
Montezuma Gun #503
Montezuma Ranch #908
Morrow Island #702
Mrs. Murphy's #128
Mulberry Land Co #705
New Family #422
North End #125
Overlook Club #322
Payton Place #505
Pickleweed Ponds #302
Piedmont #522/523
Pintail Ranch #625
Potrero #303
Rich Island #802
Riverside #929
Roaring River #906
San Souci Land #901
Sheldrake #133
Sheriff Pond #632
Shurshot #608
Sleeping Pintail #611
Spinner Island #940
Spoonbill #927
Sprigateal #413
Sprigfarm #804
Sprigsville Ranch #529
St Germaine #803
Stolte Farms #317/319
Stoned Duck #621
Suisun Farms #112
Sunrise #405
Sweetwater Gun #105
Teal #406
Tip End #426/512
Tree Sl. Farms #514
Tule Belle 416/417/421
Tule Farms #113
Tule Hilton #412
Tule Meadows #320
Tule Red #539
Volanti #219
Wagenet #605
Walnut Creek #123
Webfoot #926
West Wind #537
West Wind Sports # 508
West wind Duck #610
Wheeler Island #807
Whistler #126
Wild Turkey #517
Wildwing #330
Windmill #532
Windsong #118/127
Wingnuts #904
Wings Landing #211

PLEASE NOTE: The map of the marsh on the following pages is topographically the best quality in a small format that we have found, but it dates from 1974, and not all the property lines are identical to today's.

The general club locations may be determined by their hundred-series numbers, as indicated on the small-scale map below.

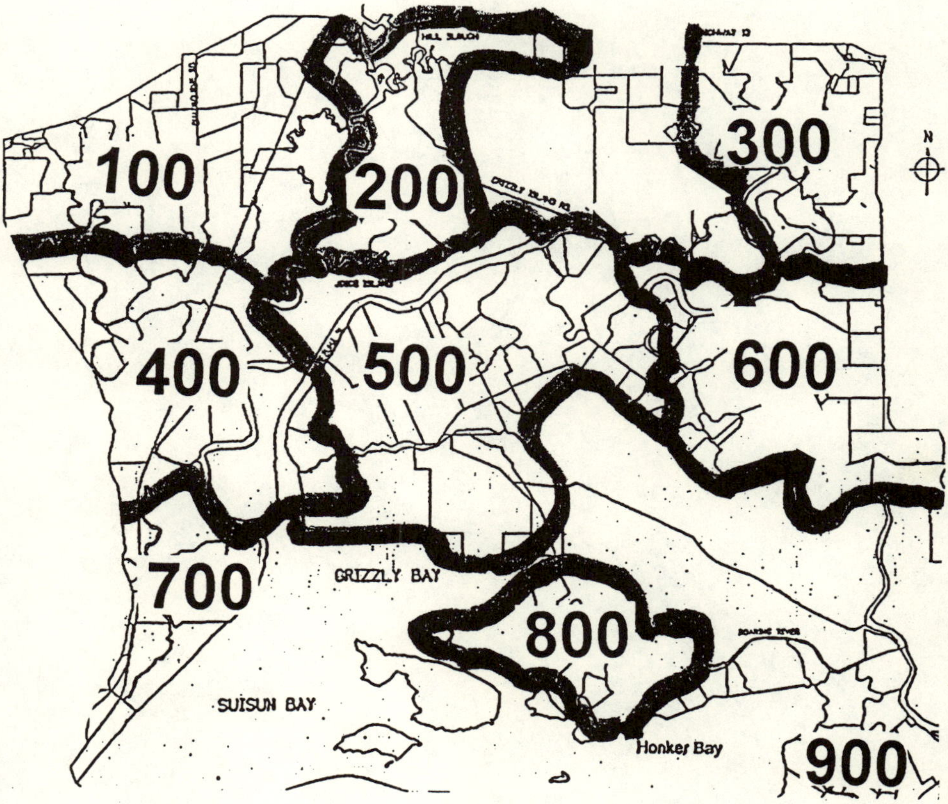

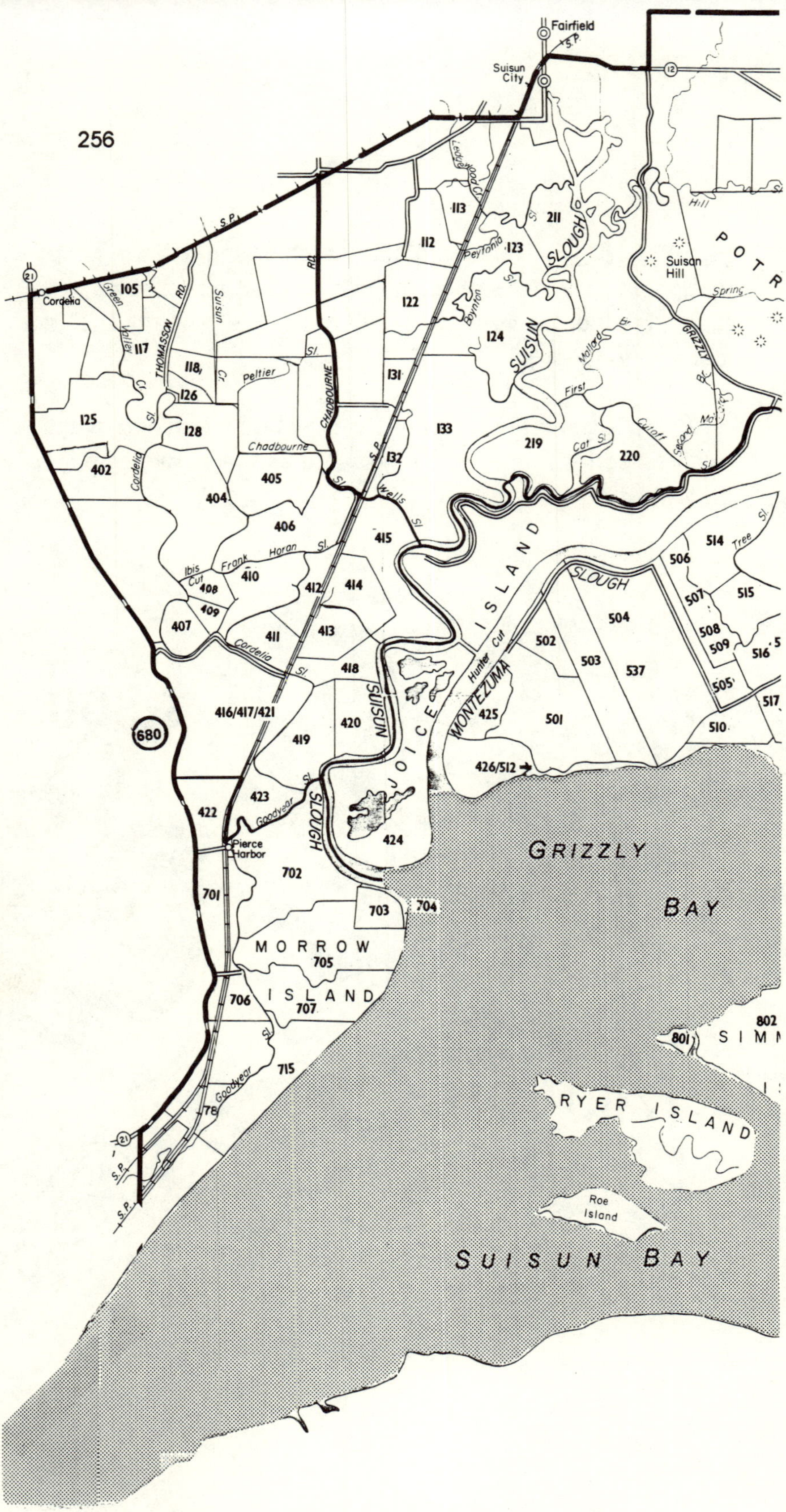
Fairfield
S.P.
Suisun City
12
21
Cordelia
105
Green Valley Cr.
117
THOMASSON RD.
Suisun Cr.
118
126
128
125
402
Cordelia Sl.
RD.
CHADBOURNE
Peltier
Sl.
Chadbourne
Ledgewood Cr.
113
112
Peytonia Sl.
123
211
SUISUN SLOUGH
Boynton Sl.
122
124
131
133
132
S.P.
405
404
406
Wells Sl.
415
Frank Horan Sl.
Ibis Cut
408
410
412
414
409
407
411
413
Cordelia Sl.
418
416/417/421
680
420
419
SUISUN SLOUGH
423
Goodyear Sl.
422
Pierce Harbor
702
701
703
704
MORROW ISLAND
705
706
707
715
Goodyear Sl.
78
S.P.
21
JOICE ISLAND
424
Hunter Cut
MONTEZUMA SLOUGH
425
426/512
Hill
Suisun Hill
POTR
Spring
GRIZZLY
First Mallard Br.
Cutoff
Second Mallard Sl.
Cat Sl.
219
220
514
Tree Sl.
506
507
515
504
502
503
537
508
509
516
505
517
501
510
GRIZZLY BAY
802
801
SIMM
RYER ISLAND
Roe Island
SUISUN BAY

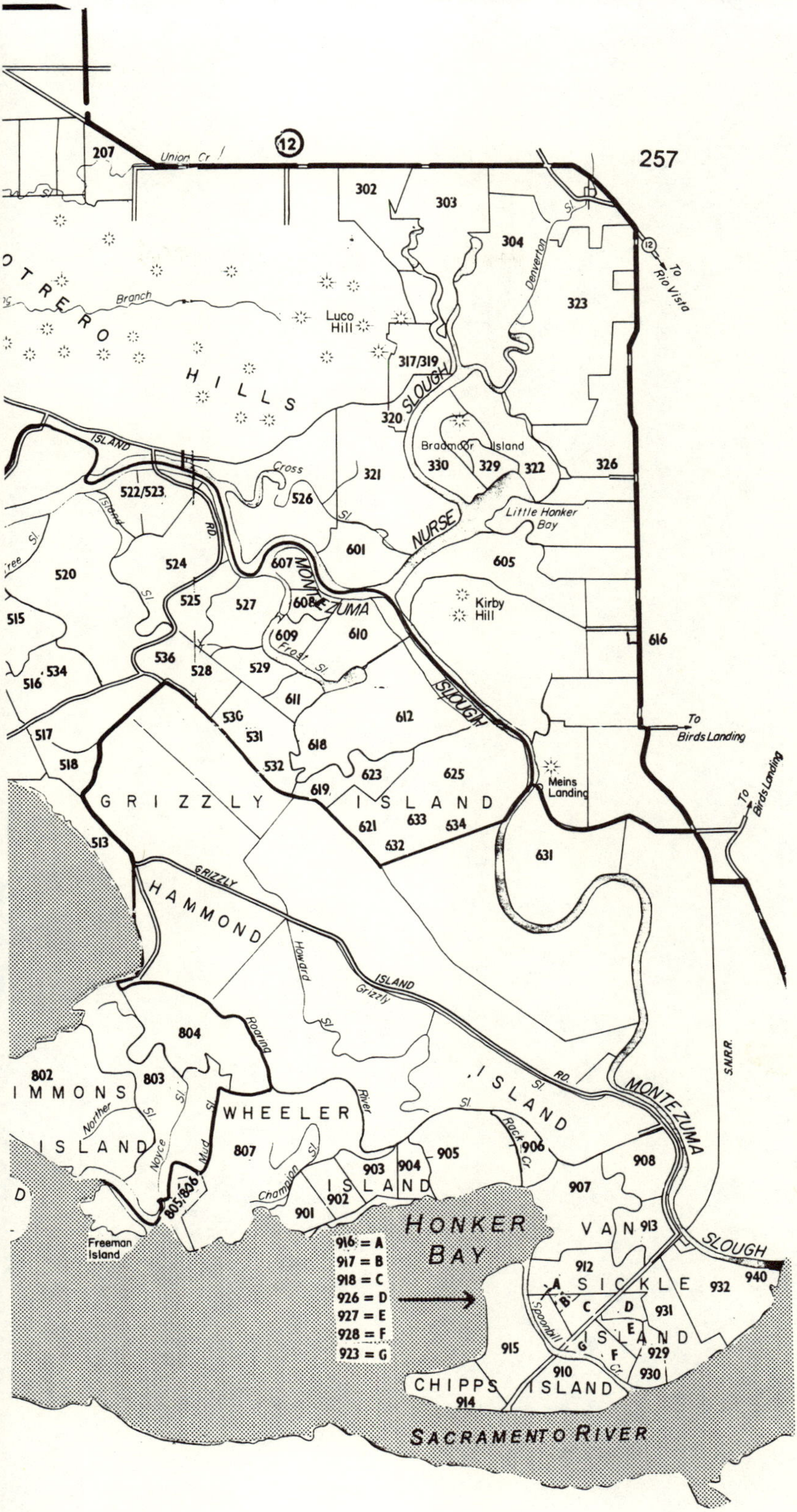

12
Union Cr.
207
302
303
304
Denverton Sl.
323
To Rio Vista
POTRERO HILLS
Branch
Luco Hill
317/319
SLOUGH
320
Broadmoor Island
330
329
322
326
ISLAND
Cross
522/523
Island
RD.
321
526
Sl.
NURSE
Little Honker Bay
601
605
520
524
607
MONTEZUMA
608
Kirby Hill
515
525
527
609
610
616
Frost Sl.
536
528
529
516
534
611
SLOUGH
612
To Birds Landing
530
531
517
618
532
518
623
625
Meins Landing
619
GRIZZLY
ISLAND
621
633
634
632
513
631
To Birds Landing
GRIZZLY
HAMMOND
Howard
ISLAND
Grizzly
Sl.
804
Roaring
802
803
SIMMONS
ISLAND
Norther
Sl.
Noyce
Mud Sl.
WHEELER
River
RD.
ISLAND
Sl.
Rock Cr.
906
905
904
903
807
MONTEZUMA
S.N.R.R.
908
907
Champion Sl.
ISLAND
902
901
805/806
Freeman Island
HONKER BAY
VAN
913
SLOUGH
940
916 = A
917 = B
918 = C
926 = D
927 = E
928 = F
923 = G
912
SICKLE
932
Spoonbill
931
ISLAND
929
915
930
910
CHIPPS
ISLAND
914
SACRAMENTO RIVER